D0845834

The Culture of Wilderness

Agriculture

as Colonization

in the American

West

The Culture of

Studies in Rural Culture

Jack Temple Kirby, editor

The University of

North Carolina Press

Chapel Hill and London

Frieda Knobloch

Wilderness

Manufactured in the
United States of America

Library of Congress
Cataloging-in-Publication Data
Knobloch, Frieda. The culture of
wilderness : agriculture as colonization in the American West /
by Frieda Knobloch.
p. cm. — (Studies in rural culture) Includes bibliographical
references (p.) and index.
ISBN 0-8078-2280-9 (cloth : alk.
paper).—ISBN 0-8078-4585-X
(pbk.: alk. paper)
1. Agriculture—West (U.S.)—
History. 2. West (U.S.)—Civilization. I. Title. II. Series.
S441.K57 1996 95-50148
306.3′49′0978—dc20 CIP

The paper in this book meets the
guidelines for permanence and
durability of the Committee on
Production Guidelines for Book
Longevity of the Council on
Library Resources.

00 99 98 97 96 5 4 3 2 1

For G.Z. and M.C.

Contents

Preface ix

Acknowledgments xiii

Introduction | Abduction.
Capturing a Poetics of
Agricultural History 1

1 Trees 17

2 Plows 49

3 Grass 79

4 Weeds 113

Epilogue.
Just the Facts, Ma'am 147

Notes 155

Bibliography 181

Index 195

Preface

This book in part describes the nature and scale of a problem of which "agriculture in the American West" is only one expression. Although the primary documents I refer to in the chapters that follow point to the West, to the period between the Homestead Act of 1862 and World War II, to the ambitions and follies of the U.S. Department of Agriculture (USDA), to a multitude of western objects, I mobilize the evidence I have collected not only in the service of "new western history" but more importantly in the service of—in anticipation of—a postwestern history. Postwestern: as in "United States out of North America," a particularly succinct indigenist, anti-imperialist, and antistatist demand, for which no "West" as such, cultural or geographical, exists.

I can't place my work in the context of having grown up in the West, intimate with the local landscape, tutored in the flora and fauna, the weather, rural work, the relationship to the "East." Nevertheless, this book came from somewhere. I had the fortune to stumble into a formative experience living and working as a ranch wife in south-central Montana for four years in my early twenties. What propelled me into graduate school and another life was a series of events like this one: a man sprayed an eighty-acre field with a powerful herbicide hoping to kill the Canada thistle that grew there abundantly. He managed to sterilize the field so that nothing could grow there for five years—nothing, that is, except Canada thistle, which spread like crazy. The thistles claimed that field definitively until he sprayed it again. Ten years after the first assault (so I was told later), a thin stand of barley signaled his victory over thistles.

The absurdity of deliberately using a biocide to "sterilize" living soil is a starting point for me and my work, not an ending point to be argued toward. The absurdity of a field covered with Canada thistle in the first place (*Cirsium arvense*, a weed of the fields since before the time of Linnaeus, bound to creep in wherever one would prefer grains), overtaken by eighty acres of barley, which would be sold to buy, among other things, groceries at a second-rate supermarket forty miles away, struck me as a sign of something deeply, violently irrational, even ruinous. "Bad farming" or "capitalist alienation," as named phenomena, didn't seem to offer any explanation.

Plenty similar manifestations demanded explanation, including how a particularly scrappy corner of town, cluttered with trailers, old trucks and parts, machines grown over with grass, came to be known as "Indian corner," the denizens of which were, not surprisingly, white. This in a place where the Absaroka Mountains, named for the people who held this land (and provided scouts, no less, for the U.S. Cavalry), break across the southern horizon, permanent and imposing. Was it the exuberant weeds, the violation of bourgeois aesthetic order, or the economic and social marginality that gave "Indian corner" its name? Other unsettling evidence of frontier malaise came from a book typical of homestead literature, Nannie Alderson's memoir, *A Bride Goes West*. The book was given to me because, like Nannie, I had packed my grandmother's china and gone west to live as a tenderfoot in the rugged world of Anglo cattle and bad prices. Nannie seemed extraordinarily unhappy to me in ways that had nothing to do with the fact that she had no running water or electricity. Indeed, her depression, social isolation, fear, her muted impatience with patriarchy, all seemed to confirm what I had begun to suspect: that the frontier was neither an isolated time nor an isolated place but, as Gloria Anzaldúa has described it, a wound kept open indefinitely, where land, people, and animals are subjected to continual cultivation on behalf of economic and cultural "progress."

Whereas these suggestive images and anecdotes broke into my life in the first person voice at a difficult time, this book takes up the relationship between cultivation and subjugation more systematically. The story of a wild nature, discovered, domesticated, and transformed into something (or someone) "productive" and "improved," is told again and again by many people, including—for our purposes here—American historiographers and agricultural scientists engaged in explaining and improving the American West. Because this story is understood as inevitable, indeed natural, and domestication and "improvement" are understood as salutary goals, the violence of such social and environmental transformations—as well as the possibility that they are unwanted, unnecessary, or at least susceptible to critique—is erased. This book offers a critique of the naturalized story of nature as becoming-culture by looking at how western agriculture and its sciences have described and transformed western people, animals, and landscapes and what has been lost or endangered in the process.

The primary purpose of such a study is not to catalog the abuses of science or agricultural practice but to draw attention to the material as well as ideological power of a naturalized history so that it can be challenged by remedial (as well as oppositional) histories. This book uses the objects and discourses of western agriculture to introduce more general questions in the spirit of such a history: Re-examining the history of agriculture, how can we imagine uncoupling state formation from food production, the technologies of war from the life-sustaining practices of farming, the work of farmers from the drudgery of commodity production? How can we take back agricultural expertise from agricultural science—that is, how do we democratize the knowledge as well as the practice of food production? How can the principle of usufruct be reinvented and applied to food production in North America late in the twentieth century? How can we ensure that our answers honor indigenous land tenure and self-determination as well as social, economic, and environmental democracy all around? These questions were inchoate when I first made the decision to write about the West; they emerged with greater definition as I examined the evidence of western social and environmental history and have ultimately changed how I think about history, work, land, food, and people well outside the West.

Acknowledgments

This book's debts are many; what I owe exceeds what it is possible to say in the "acknowledgments" genre. All of these people have my sincere gratitude for making many important things possible: Nancy Armstrong, Joe Austin, Dan Barclay, Bill Bevis, Rachel Buff, Maria Damon, Susanne Dietzel, Louise Edwards, Nan Enstad, Jim Farrell, Lisa Fischman, Rich Kees, Jack Temple Kirby, Frank Knobloch, Susan Kollin, Carrie Krasnow, Gretchen Legler, Vicky Munro, David W. Noble, Constance Pierce, Riv-Ellen Prell, Guy Puccio, Joanne Puccio, Paula Rabinowitz, Tony Smith, Gordon Teskey, and Van Zimmerman.

I am also grateful to the officials of the state departments of agriculture and the many county weed inspectors with whom I spoke and corresponded during the summer of 1991 as I prepared a report on the effects of weed control laws on weed acreage. Through that project, I collected many documents that became primary resources and inspiration for my chapter on weeds. I owe special thanks to William Scott of the Kansas State Board of Agriculture; Jack Peterson of the North Dakota Department of Agriculture; Dennis Clarke of the South Dakota Department of Agriculture; Kenny Rogers, county supervisor of Yuma County, Colorado; Geir Friisoe of the Nebraska Department of Agriculture; Dale Clark of the California Department of Food and Agriculture; Randy Westbrooks of the USDA Animal and Plant Health Inspection Service; and Buck Waters of the Bureau of Land Management. Thanks also to Frank Forcella of the USDA Agricultural Research Service at the North Central Soil Conservation Research Lab, Morris, Minnesota, for copies of the weed introduction and migration reports he produced with Stephen Harvey. I thank weed ecologist Bruce Maxwell for giving me the chance to work with him.

A Norman Johnston Dewitt Fellowship at the University of Minnesota made the completion of the first draft of the manuscript possible. The reference staffs at the Anderson Horticultural Library and St. Paul Central Library (both at the University of Minnesota) provided expert assistance.

At the University of North Carolina Press, Lewis Bateman was an engaging, clear-minded editor. Two anonymous readers provided insightful comments on the manuscript. I also thank Kathleen Ketterman and Paula Wald at the Press for their exceptional good sense.

Finally, Grete Zimmerman was both patient and curious about a project that could only have appeared mysteriously interminable. She weathered a lot of it with me in rounds of library visits, haphazard meals, breaks for stories, and hours of disjointed conversation as she played and I "composed." Michael Corbin was a stalwart co-conspirator and always asked the hardest questions. I hope he never stops asking them. Whatever good comes of this book is for these two.

The Culture of Wilderness

Introduction | Abduction. Capturing a Poetics of Agricultural History

> All natural science is nothing but an attempt to understand man and what is anthropological; more correctly, it is an attempt to return continuously to man via the longest and most roundabout ways. Man's inflation to a macrocosm in order to be able to say in the end: "in the end you are what you are."
>
> —Friedrich Nietzsche, *Philosophy and Truth*

Colonization is an agricultural act. It is also an agricultural idea. Both the material fact and the idea of colonial agriculture took place in the American West—that is, on the advancing line of Euro-American settlement in a geographical place west of the Mississippi. By common usage, the vast colonial outback that Jefferson secured in 1803 for an agrarian empire is the most likely location for any study of "the West." This study is no exception. But by asking how agriculture developed in that place after 1862, I am really only amplifying a difficult and less localized question: How exactly do agriculture and colonization work together as ideological and material practices? Exploring how western agricultural conquest worked can help us understand the extent to which the conquest of "the West" informs every plowed field and who we are as a society still fed (materially and otherwise) by the agriculture perfected after a fashion in the West by 1945.

As a physical place and as the moving frontier, the West was assumed to be wild in the Euro-American imagination. A wilderness inhabited by savages was overtaken by people barely tethered to the institutions of civilization who would make this natural domain the cradle of a distinctly American culture—so the story begins. Many of the faults of this story are laid bare in what has come to be called the "new western history." The part of the story that concerns me in this study are two related historiographical tropes: the present emerging, as if inevitably, from the past, just as culture emerges inevitably over time from nature. This inevitability is naturalized. It cannot *not* happen that something primitive, simple, and primordial (associated with the past) gives way to something complex and sophisticated, just as it cannot not happen that an organism grows and develops or a child matures. Hence the association of childishness with all things assumed to be found in a "state of nature," such as indigenous populations. Landscapes understood to be natural are likewise understood not only to be untouched but also to be *waiting* for civilizing instruments to develop them, as if that is their inevitable fate: a virgin prairie ready for the plow.

Destiny is indeed manifest in a world where nature exists only to give way to culture and the past develops organically into the present. What is not manifest—what is obscured—in such a world is the force of will engaged in this project, a social will rather than a natural(ized) developmental force that informs tools, laws, ambitions, knowledge gathering, wars of conquest, preferred forms of territorial occupation. This social will, a will to conquest that predicates itself on its naturalization, is what I describe in the four chapters here. My goal is partly to denaturalize the West and the frontier, to sort out what social agendas have been at work in western agricultural history. This aspect of the book might be described as a new western historian's revision of once-celebrated (in many places still celebrated) triumphs of western agriculture: cleared forests, prairies under the plow, fenced range, the increasingly high-tech war against weeds.

But the philosophical, historiographical problem of nature and culture, and of the emergence of the present, is as central to this study as the western objects that are its most immediate subject. My analysis of western agricultural practices and sciences is not just a reexamination of the past from the present but an interrogation of a domain of scientific expertise from the domain of the humanities. These chapters are intended to suggest a line of critique of agriculture in general as composed of those forms of expertise, land use, tools, and social relationships that systematically (and at great social and environmental cost) turn nature into culture. The purpose of such a critique is to render the epistemological habits that feed us *accountable* to us, socially, politically, materially, technologically.

Agriculture as such has never simply been about raising food crops or the sciences that make this more productive and efficient. Agriculture is an intensely social enterprise, shaped by inescapably social desires and expectations, even if it is described in simplistic material or natural terms: working the land, improving a breed. What is implied in these formulae is a whole system of domestication—that is, the transformation and improvement of nature—that is as much about structuring social and political life as it is about raising cattle or wheat. This is why agriculture ought to capture the attention of political activists and theorists, cultural studies scholars, intellectual historians—*not* just because we all eat (which is a limited but certainly legitimate reason to concern oneself with where one's food comes from) but also because the cultural work of agriculture is so central to many of the problems progressive activists and theorists work on. The West, then, and the agriculture that domesticated it, becomes a specific embodiment of cultivated nature, a test case. There, the categories of nature and culture and the history they imply are particularly legible because of the West's continuing identification with wilderness and the significance of the frontier in American historiography.

The first chapter, "Trees," examines the forest as the origin of civilization in European understandings of cultural progress—cut down to clear land for fields—and the subsequent transformation of trees into agricultural objects to be managed as a crop themselves. The history of western logging and timber conservation provides an excellent example of how agriculturalization (turning trees into crops) involves the imposition of state power and institutional knowledge production, deployed by ranks of official experts and even military agendas, on western territory and objects. The consolidation of military and agricultural power over western forests was complete by the end of World War II, after which foresters systematically turned their attention to the "development" of forest resources abroad, especially in the Third World.

"Plows" presents the history of the moldboard plow, its improvements to benefit prairie farmers, the scale and value of western land clearing, as well as some of the unexpected results of western crop production. Here I describe agriculture as a philosophy based on the categories of nature and culture, defined by the relationship between the land (understood as wild and unproductive) and the domesticating, improving work of the plow. The plow is axiomatic of agriculture as such because it is so intimately a sign of civilization to Europeans and Euro-Americans. Moreover, because the plow has historically been a tool of men, having displaced hand tools and the predominantly female farmers who used them, the conquest of the "virgin" prairie under the plow is deeply marked by gender, not only ideologically but materially as well. The produce, tools, chemicals, and social relations institutionalized by western agriculture were exported after World War II—first in "food aid" to Europe immediately following the war, then in the Green Revolution to parts of the Third World.

The third chapter, "Grass," traces the philosophy of nature-as-becoming-culture in the history of the western range, an area largely untouched by the plow. Rather than imposing civilization on the wild landscape by plowing it, western ranchers sought to improve nature by breeding up their wild or scrub livestock, improving their pastures by seeding what they referred to as tame grasses, and managing their stock's grazing patterns. Eventually, the range became an agricultural entity managed for the maximum productivity of its crops (hay, forage, and high-quality stock) without ever turning a furrow.

The fourth and last chapter, "Weeds," describes the floral demimonde of unwanted plant pests as the irrepressible margin of western agriculture, thriving in spite of every effort of eradication. I examine the anthropomorphic qualities of weeds in the literature on weed control and argue that the persistence of weeds signals the need to establish a more socially and economically as well as biologically diverse agriculture in the West and elsewhere.

An epilogue addresses the problem of historical writing and the archival sources it draws upon in the creation of a limiting view of the past. It argues finally for a reclamation of the knowledge and practice of food production from colonial, agri/cultural experts.

The word "agriculture" was recorded in English for the first time in 1603. The word "colony" entered the English language in 1566; the verb "colonize" appeared in 1622; and the process of colonization had a name in 1770.[1] The ideas and practices of agriculture and colonization that are my subject, then, are of relatively recent origin. I take "agriculture" to mean agriculture *as such*—the culture of food production that gave rise to the word "agriculture" in 1603, not a generic designator for how any society produces food or even how Europeans produced food a century or two before that. Euro-Americans have tended not to see indigenous food production as agriculture at all, indicating the cultural specificity of this word. To wish that white settlers had been able to extend their definition of agriculture to include, say, wild rice gathering, is to miss an opportunity to understand language as having a history that belies the "obviousness" of ordinary words like "agriculture."[2]

Agriculture as such is "the science and art of cultivating the soil." Cultivating, in turn (another seventeenth-century word), means to put labor into improving the land by tilling it. Agriculture—the culture of the fields—is inherently about culture as art and science (certain kinds of labor), and changing the land by "improving" it, not simply about food production. The same attention should be given to the word "colonization." Of course, we use it often to designate conquest by force and the exploitation of resources, and many civilizations have perpetrated such conquests. The word "colony," however, was derived specifically from the Latin word for farmer, at a time when European landowners were colonizing their own backcountry, enforcing their ownership by bringing new lands into cultivation, changing the land-use practices of peasants, and forcing many of them off the land.[3] This was a violent and disruptive process. The two words work together: colonization is about enforcing landownership through a new, agricultural occupation of lands once used differently. Colonization is a good thing, according to its supporters, regardless of the bloodshed and disruption it creates: it brings about the "improvement" of land newly under cultivation—it brings culture to a wilderness.

In the chapters that follow, I present components of western American agriculture with an eye to the colonial purposes that inform them. Like the words "agriculture" and "colonization," the entire vocabulary of western American agri/culture has a history of its own as well as being part of the history of the American West. By necessity, this is a textual study. I have access to the forms western American agriculture has taken through the USDA documents, journal articles, agricultural textbooks, critiques, and memoirs written about it. Even the material effects of this agriculture—how colonization has been inscribed on the body of the continent—have to be treated as textual matters, since even if we can see them, I can't show them to you here. But we should not assume that the textual evidence is an inferior substitute for "real" knowledge about the material practices and effects of western agriculture. The two are intimately related, and it is part of the work of this study to trace the relationships between agricultural ideas (a vocabulary) and material agricultural practices and their effects (as they have been recorded). This study might be called a grammar of western agriculture, indeed a poetics of western agriculture—a grammar to describe the "rules" of western agriculture, a poetics to show how these rules refer to phenomena outside western agriculture.

This study claims several theoretical models and scholarly antecedents. Perhaps the most obviously direct of these is Donald Worster's sustained critical inquiry into the agricultural and environmental history of the American West, to which I owe a considerable debt. However, my attention to the language and images of western agriculture is more closely related to the work of American studies scholars, notably Henry Nash Smith and Annette Kolodny, who—in contrast to Worster's materialist focus—address what they call the myths, symbols, and fantasies that accompanied white agricultural expansion in the West. Finally, this study is informed by the arguments of theorists of language and philosophers of history, including Jacques Derrida's (and Friedrich Nietzsche's) understanding of objective truth as metaphoric in origin; Carolyn Merchant's practice of "ecological history" as including ideology, technology, scientific ideas about nature, social history, and conceivably any other category of the human environment; and Michel Foucault's practice of "archaeology," as opposed to intellectual history, of the human sciences, whereby disparate sciences are read for the *epistemes* they nevertheless share. What is at stake here is at once an evaluation of western agricultural history as well as a distinct historiographical trope—that something called culture emerges from a primordial thing called nature in predictable and desirable forms.

Worster's theory of history includes what he sees as a necessary reintroduction of nature into human historiography. Therefore, although he writes environmental history, he does not do so as a practitioner of a subdiscipline but with the sense that environmental history has something to teach historians as a whole about humans' dependence on nature and the limitations of human thinking—however powerful—among all the forces of nature. Hence his titles, "Transformations of the Earth: Toward an Agroecological Perspective in History," not just environmental history, and "History as Natural History."[4] Worster's historical practice involves three levels of inquiry and choices of historical subjects. The first level is nature, the historical study of which involves "the discovery of the structure and distribution of natural environments of the past," based on an understanding of "nature" itself through the use of the work of natural scientists. The second level is the interaction of human societies with nature through their "modes of production," that is, through their "tools, work, and [resulting] social relations." The historical study of this level is "more fully the responsibility of the historian and other students of society" rather than natural scientists. It is concerned with the "various ways people have tried to make nature over into a system that produces resources for their consumption." The third level "is that more intangible, purely mental type of encounter [with nature] in which perceptions, ideologies, ethics, laws, and myths have become part of an individual's or group's dialogue with nature." He does not identify which scholars might be best suited for the study of this level, although elsewhere he notes the contributions of the Frankfurt School in furthering the understanding of the ideological human-nature dialogue.[5] But even in this realm, the historian's subject is a dialogue with nature, not the (human) terms of the dialogue itself. Worster focuses on human relationships with nature rather than relationships among humans themselves. He follows the Frankfurt School in concluding that "human domination derives from the incessant modern drive to remake nature," implying that a right relationship with nature would make humans' domination of one another either less likely or impossible.[6]

Worster concedes that the "lines of historical causality" run in both directions through the three levels of history he identifies and concludes that "most environmental historians" are both materialists and idealists, without a single theory of causality.[7] But Worster's preference is decidedly, even emphatically, materialist: "Apparently, some of my commentators . . . feel I am underestimating the centrality and force of humans in environmental history; that is, I am too much of a determinist, materialist, naturist. Good: at least I am understood in my tendencies."[8] Walter Prescott Webb and John Wesley Powell are his precursors in a materialist line, "steadfast, clearminded regionalists" who had the sense to define the West in concrete geographical and climatological terms.[9] Worster sees himself, Webb, and Powell as contributing to a western history at odds with that of Frederick Jackson Turner, who was preoccupied with an abstract, elusive process of the "frontier" rather than a place called the West. Turner "started historians down a muddy, slippery road that ultimately leads to a swamp" of generalizations and abstractions rather than offering the "concrete promise" of explaining American history by reference to the process of settling frontiers.[10] Worster's subject is the real West, not an abstract, processual West, and he takes his definition from Webb and Powell as that place beyond the hundredth (or ninety-eighth) meridian.

The work of Henry Nash Smith and Annette Kolodny represents a very different approach to the history (agricultural and otherwise) of the American West. Henry Nash Smith drew a great deal of attention to cultural history (not to mention the question of what American studies was, anyway) with the publication of *Virgin Land: The American West as Symbol and Myth* in 1950. Smith's subjects in this book were the intellectual constructions and images that had "exerted a decided influence on practical affairs" in the westward settlement of the United States by Euro-Americans.[11] What he called the myth of the garden was preeminent among these. Thinking that the continent was a paradise regained, farmers and statesmen alike behaved as if the "virgin land" of the continental interior could be brought under glorious cultivation, to the virtuous enrichment of individual farmers as well as the republic at large. The Homestead Act, the development of railways, and the establishment of land-grant agricultural colleges and the U.S. Department of Agriculture were all justified by this assumption.

Of course, the hundredth meridian (that stern limit of Worster's West, as well as that of Walter Prescott Webb and John Wesley Powell before him) presented an obstacle to untrammeled expansion, but some people were confident that even the desert beyond that fateful line could be made to blossom as the rose. Josiah Strong, a booster for western expansion and the racial superiority of Anglo-Americans, wrote in 1885: "The rainfall seems to be increasing with the cultivation of the soil."[12] It was exactly this imaginative capability that, for Smith, made history happen; it is exactly this sort of ideational arrogance that Worster deplores. For Smith, images like a desert blooming as a rose "are never, of course, exact reproductions of the physical and social environments"—a materialist concession—but "history cannot happen—that is men cannot engage in purposive group behavior—without images which simultaneously express collective desires and impose coherence on the infinitely varied data of experience."[13] With symbols and myths as his subject, Smith chose to traffic in the cultural realm of history rather exclusively. He explicitly remarked that he was not interested in whether "such products of the imagination accurately reflect empirical fact," although he understood a "myth" to be something other than just an "erroneous belief" in its impact on human history.[14]

Annette Kolodny, a former student of Smith's, begins with the same assumption in her work. Like Smith, she is a reader of textual records of western settlement, attentive to what they say about the settlers' attitudes toward their work and environment. Kolodny is especially interested in the differences between white men's and women's ideas about the West because the former dominated western settlement with the imposition of fantasies of exploitation and large-scale transformation. She wrote, "Given the choice, I would have had women's fantasies take the nation west rather than the psychosexual dramas of men intent on possessing a virgin continent." In contrast, according to Kolodny, women wrote about the West as a garden to be tended, and this garden "implied home and community, not privatized erotic mastery."[15] Although she alludes to the relationship between ideas of dominance and exploitation and their actual expression in a material project of conquest, Kolodny is not preoccupied with proving that the former influenced the latter or with defining an idealist theoretical argument. She states briefly in her preface to *The Lay of the Land*: "That the symbolization [of the land-as-woman in American life and letters] appears to have had important consequences for both our history and our literature should not suggest, however, that it accounts for everything, or that to it, alone, must we attribute all our current ecological and environmental ills. . . . At best I am examining here only a link in a much larger and much more complex whole; but it is a vital and, in some cases, a structuring link—and one that has been for too long ignored."[16] Kolodny, like Smith, is at home taking cultural constructions (myths and symbols, or fantasies) as important historical subjects in their own right, without dwelling on exactly how a system of symbols structures the behavior of the people who use them.

This approach, especially as practiced by Smith, has come under fire for ignoring "real" (material) history, for ascribing universal properties to a culturally specific (even class- and gender-specific) set of myths and symbols. Smith's book was later seen to be paradigmatic of an entire "school" of American historiography that presented idealist arguments which could not be proved by any reference to concrete data and indeed seemed to look down upon the burden of proof carried by more "legitimate" historians, among its many faults. The first generation of scholars working in American studies, as self-consciously and institutionally distinct from either history or literary studies, was later called the myth and symbol school by Bruce Kuklick in 1972, after the subtitle of Smith's seminal book marking the emergence of a new historiographical moment in the practice of American studies in the 1950s.[17] Literary New Criticism had arrived on the academic scene by that time, bringing a whole flock of words for imaginative constructions, including "archetype" and "image," along with "symbol" and "myth." Indeed, in literary studies, this type of analysis was understood to be rather objective and scientific (unlike literary appreciation), as underscored by the title of Northrop Frye's *Anatomy of Criticism*. This emerging critical style in literary studies had collided (perhaps unreflectively) with historical study in the work of scholars whose concerns encompassed both literature and historical research. One could read history the way one read poetry, with a close attention to images. One could also ground literary productions in a historical context—something the literary New Critics were accused of avoiding. Many practitioners of American studies active between 1940 and 1965, including Smith, Leo Marx, Perry Miller, R. W. B. Lewis, and others, read literary texts along with historical documents—belles lettres—and the prey they stalked through those promiscuous pages were ideas, not things.[18]

Kuklick notwithstanding, some scholars of American history have continued to understand symbols and myths (along with ideas, ideologies, images, and more differentiated vocabularies and metaphors) as legitimate objects of inquiry, and a few of these—including Richard Drinnon and Richard Slotkin, along with Annette Kolodny—are scholars of the American West. Not surprisingly, given Donald Worster's appraisal of Turner, these scholars find something valuable in Turner's work: not his history as the unqualified "truth," certainly, but his language, and the ideological conventions he captured in his rhetoric, which they read as symptomatic of a different kind of historical truth—evidence of the power and pervasiveness of certain ideas that accompanied westward expansion. William Cronon notes that within thirty years of his death, Turner had few defenders as a historian but "the master was now studied more for his rhetoric and ideology than for his contributions to historical knowledge."[19] For Slotkin, Turner described a conceptual world (bifurcated between Wilderness and Metropolis) that had existed in the popular realm apart from Turner since at least 1620. For Drinnon, the "hauntingly evocative lines" of Turner's famous 1893 address were effective and influential because he alluded to "conventional wisdom built up by three centuries of expansion." Kolodny cites Turner's characterization of the continent as providential Mother as one of many examples of men's exploitative fantasies of the West.[20] Turner merely gave virtuosic expression to these ideas—they were not in any originary sense his—and the ideas themselves (the female continent, the western garden, the American desert, the American character), however "fictitious," became the bread and butter of idealist scholars of American history.

Like Worster, American studies scholars have largely formulated their ideological explications without taking on the question of how the ideal and the real relate to one another conceptually or providing any substantial theoretical account of the significance of ideological language. Worster at least outlines a theoretical model, via the Frankfurt School, for how culture and nature interact with one another in human modes of production, even if he does not question the categories of nature and culture themselves. After sustaining criticism for having failed to justify an idealist method, both Leo Marx and Henry Nash Smith later (in the 1980s) cited cultural anthropologists like Clifford Geertz and Marxist literary scholar Fredric Jameson as theoretical allies in projects these writers had taken on much earlier in comparative theoretical silence.[21] Smith merely listed three surnames of people who had influenced his thinking in his 1970 preface to *Virgin Land*: Vaihinger, Bergson, and Lévy-Bruhl—two philosophers and an anthropologist.[22] These names are all suggestive of fictions (Vaihinger was explicitly a philosopher of fictions): the fiction of rational science that allows the human will to impose its ideas on nature and other people (Bergson and Vaihinger) and the fictional conflation of real people with the symbols they create to represent themselves (Lévy-Bruhl).[23] Vaihinger's study of fictions is in fact a synopsis of European ideas regarding truth, reality, and nature, running through Kant, Darwin, Hegel, Schiller, and Schopenhauer and closing with a long exegesis of Nietzsche. For Vaihinger, "truth" mediated between a set of ultimately unknowable material contingencies in order to gain some practical, if limited, control over them. Science did not reveal truth so much as it set it up like a scaffold to deal with nature. Smith believed he had lost sight of this notion of truth by the end of *Virgin Land*, a lapse he felt compelled to address in the 1970 preface. "I was trying to make a valid point," he wrote, "I wanted to protest against the common usage of the term 'myth' to mean simply an erroneous belief... [but nevertheless] I tended to conceive of [myths] simply as distortions of empirical fact."[24] But Smith (like Kolodny, Slotkin, and Drinnon) saw his work first as American historiography and second, if at all, as theoretical exposition. By failing to do more than list his influences, he cut himself off from the possibility of incorporating the influential ideas into a more systematic presentation of the relationship between myths and symbols and the American West. Worster's materialist model is more complete and rhetorically more convincing because Smith and other scholars did not use some of the most powerful theoretical tools at their disposal to interrogate the issue of "empirical fact" and its relationship to imaginative constructions.

Perhaps the most important "empirical fact" of contemporary environmental history (indeed history in general, following Worster's excellent recommendation) is the fact of nature—its existence and its agency. Carolyn Merchant's work goes a long way toward understanding that this most basic fact, the one against which we learn to measure what culture is and what it has historically accomplished, has its own history—not natural history as studied by biologists or geologists but an intellectual and discursive history. *The Death of Nature* describes how our understandings of nature—even, or especially, scientific ones—

have changed over time. Like American studies scholars Smith and Kolodny, Merchant is closely attentive to language. She reads scientific texts of the past for the vocabulary with which they describe nature and for how this vocabulary changed over time. Up to the sixteenth century, nature was understood as a beneficent female, a living being who supported all life. This view changed over the course of the next two centuries—accompanying changes in the social, material, and economic lives of Europeans—until nature became something dead and mechanical, clocklike and perfect, perhaps, but something to be operated rather than honored as an active principle in its own right.

Merchant does not get bogged down in this book explaining the lines of causation between the scientific revolution, its construction of nature, and the rise of capitalism and reorganization of European societies; she assumes that all of these issues can only be understood if taken together in an "ecological perspective," addressing "a broad synthesis of both the natural and cultural environments of Western society at the historical turning point" of the advent of modern science. Causation is not her subject. The many changes concomitant with the scientific revolution *together* "resulted in the death of nature as a living being and the accelerating exploitation of both human and natural resources in the name of culture and progress." She notes that the idea of "reality as a machine rather than a living organism, sanctioned the domination of both nature and women."[25] But this sanction is not a determinism. It is part of a much larger phenomenon whose complexity renders the search for the direction of causation beside the point. Merchant's ecological point of view, while it may seem to abdicate the historian's charge to understand why history happened as it did, fulfills perhaps a more important obligation: she resists leaving out those pieces of the past whose omission obscures the domination of one group by another or the fact that one kind of domination (of something understood as "nature") accompanied another (of women). Causation is not so much the point here as are the variety of forms of domination practiced by modern European, and then American, society. Merchant inquires after not the determination of history but its overdetermination.

Merchant also makes clear that nature is not a thing but an idea about which we have changed our minds. She is just as eager as Worster not to leave human effects on the nonhuman world out of her history, but she is careful about her vocabulary—perhaps because other people's vocabularies are integrally part of her subject—and "nature" is simply not a reliable term for what lies outside human (or cultural) experience. Just as importantly, Merchant presents science as an intellectual pursuit that, its modern claims to truth-telling notwithstanding, has created a specific version of nature on which its truths are based. The idea of nature-as-machine, then, born from science as simply "nature," is one of the most important ideas in the modern intellectual repertoire, all the more so because nature is forced to stand in for exactly that portion of reality outside our consciousness.

For Merchant, neither this idea of nature nor the science that purports to describe it are entirely trustworthy. She does not, however, slip the rug out from under science cavalierly. Science is part of her complicated history, one of many intellectual pursuits (historiography being another), and she notes that the mechanical paradigm that has lately defined nature (as a perfectly ordered, closed, and therefore predictable system) may be giving way to a new nature described by nonlinear ideas of transformative processes and disorder. In the mechanical paradigm, things happen because external forces act on inert objects, and the resulting motions are predictable. In the new science, things behave differently depending on the context one finds them in, and they change or create changes in unpredictable ways. Such a shift in the concept of nature would be accompanied by other shifts in the realms of politics, economics, social life, letters, and so on, and Merchant welcomes this change as a sign that nature—and therefore other things as well—will no longer be considered masterable in the old exploitative ways. A "new world view" may be coming into being "that could guide twenty-first-century citizens in an ecologically sustainable way of life."[26] This is a possibility not determined by science but augured by it among other things, including Merchant's own complex ecological historiography.

Merchant's theoretical model for this kind of history is based on the movement and exchange of materials, energy, and information throughout the amalgam of the human and nonhuman world, all as understood through an inescapably human consciousness.[27] Her ecology-based rather than nature-based cosmos defines material existence, energy, and information as exclusively proper to neither humans nor the nonhuman world. There is no nature from whom culture has become alienated or is in need of reintroduction to. Merchant's understanding of reality—both human and nonhuman—also has the advantage of addressing the issue of reproduction. Without accounting for reproduction in the human as well as nonhuman world, a historian obscures the role of women and in a more general sense obscures the conditions of (re)creating or sustaining anything that exists.

But it is still science, however qualified as a purveyor of the truth about nature, that gives Merchant her central vocabulary of energy and biogeochemical exchanges into which all the rest of reality can be translated. Even if her history is not divided into three levels, with natural science at the grounding bottom, it remains grounded, as most environmental history is, in concepts provided by scientists—that is, physicists, biologists, and other practitioners of the sciences as opposed to the humanities—about the nature of nature. Given Merchant's generous construction of historical complexity, science and its language need not be at the center of things explaining everything else. Moreover, human consciousness, if it is understood to be bound up in the ubiquitous exchange of energy and information, need not be relegated to that celestial sphere farthest from the minerals, even graphically.

Murray Bookchin's social-ecologist philosophy offers a theoretical understanding of history and reality as comprehensive as Merchant's but without her reliance on contemporary scientific concepts as an ecological core. Bookchin recovers what he understands as the reason of all reality, human and otherwise, and presents human consciousness as a natural outcome of an evolution that is at once social and biological. He, like Worster and Merchant, is interested in the relationships between forms of social organization that accompany certain technologies, different forms of reason (in Merchant's case), and competing uses of the nonhuman world. But these scholars have very different approaches: whereas Worster begins with the nonhuman world, or nature, and Merchant reveals flows of energy and information, Bookchin begins with society and its patterns of domination. For him, human society perfected technics of administration and control that made the domination and exploitation of nonhuman nature possible *and likely*. Drawing on the work of Lewis Mumford, Bookchin goes to some length to describe these technics as ideological structures, not just mechanical or narrowly technological ones. It "is difficult for us to understand that political structures can be no less technical than tools or machines," he writes, and it is easier to deal with how a specific technology achieves "certain destructive or constructive forms on the natural landscape than to explore the deformations they produce within subjectivity itself." A coercive authoritarian technics organizes the "implements, work and imagination involved in the modern technical ensemble." Bookchin's work in general represents an effort to define what he sees as a genuinely libertarian, anarchist technics.[28]

The language in which a technics is described, including the scientific discourses that inform and sustain it, is never "merely" descriptive but always performative as well; scientific discourse provides essential evidence of the multifarious work performed by a society's tools, ideas, and institutions. It is in this context that the idea of a poetics of any science—including a poetics of agriculture—is possible and relevant.

Identifying scientific theories as a form of "plot"—in which a hypothesis seeks to "organize 'events' systematically," as they might occur, or in their probable or necessary order—Fernand Hallyn justifies the use of the word "poetics" to describe a tropological study of scientific theories.[29] He argues through Alexander Koyré that scientific ideas are related to all sorts of other ideas, including explicitly literary ones; and he argues through Foucault that "profound constraints of an epoch ('epistemes') . . . lead to the production of homologous forms of discourse in seemingly unrelated domains" like science and philosophy, history and literature.[30] He reveals the "poetic structure of the world" by examining the relationships between two scientific cosmologies—those of Copernicus and Kepler—and the "various languages or signifying practices" of the culture in which these cosmologies emerged.[31] He is interested in the occasions on which hypotheses like these were created, in "the play of the imagination in an indefinable act of creativity," which has often been described by intellectual historians in nonrational or even poetic terms and which Hallyn calls "abduction."[32]

The scientific model produced by abduction is a metaphor, one kind of trope. Scientific tropes are "descriptive instruments," reliance on which "has the advantage of exposing certain analogies, certain relations [with other signifying practices], the nature of which must still be studied in depth."[33] Scientific tropes are, at the moment of their appearance, unusual; they have the "fresh and different look" of the person (the "subject") who abducted them. But the subject "aims to subjugate himself and others to the constraints that the object [the model, the abduction] is supposed to impose on all." Bad models remain tropes—recognized for their unusuality. Good models—those that can be put to instrumental use with some reliability—are "duly confirmed and accepted" as a "new *proper* language," no longer understood as metaphors or models, at a moment when "poetics gives way to epistemology and the history of science."[34] Hallyn, using Vico, Nietzsche, and Lacan, argues that a "deep tropology" informs our myths, language, and desires, an "interior tropology that shapes the general nature of our representations of reality,"[35] including scientific representations, which so easily become identified with reality itself. Carolyn Merchant's *Death of Nature* is, in effect, a tropology of modern science, but she does not use that language to describe her project, presumably because all of her work courts an easier credibility with an audience that believes in some objective scientific truth.

Other historians, notably Foucault and Hayden White, write for audiences different from Merchant's and consequently take different historiographical liberties. Foucault's *Les mots et les choses* (Words and things) was translated as *The Order of Things: An Archaeology of the Human Sciences*. In it, he sought to describe the "systems of regularities" in scientific discourse, not whether a given theory was "true" but the discursive conventions a naturalist, economist, or philologist used in creating something that had "value and practical application as scientific discourse"[36]—in other words, what rules governed representations of the truth. Otherwise unrelated sciences brought words and things together in similar ways to describe a society's conception of order.

Hayden White's tropology of history elaborated in *Metahistory* is closely related to Hallyn's search for the "plot" of science, in which each historical work is read "as what it most manifestly is: a verbal structure in the form of a narrative prose discourse."[37] The sciences of Foucault's archaeology each account for the history of certain phenomena: nature, money, language. White's metahistory is an archaeology of historiographical practice, revealing how different kinds of history account for the past as an object. Using Nietzsche, White describes the "metaphorical mode" of nineteenth-century historiography as the understanding of history as art and historical "facts" as traces of a historical will, a social desire, without which such facts or their orderly arrangement would have no meaning.

Nietzsche's essay, "On Truth and Lies in a Nonmoral Sense," directly addresses the problem of knowledge as a property of human life and society. It is in this essay that Nietzsche writes: "What then is truth? A movable host of metaphors, metonymies, and anthropomorphisms: in short, a sum of human relations which have been poetically and rhetorically intensified, transferred, and embellished, and which, after long usage, seem to a people fixed, canonical, and binding. Truths are illusions which we have forgotten are illusions; they are metaphors that have become worn out and have been drained of sensuous force, coins which have lost their embossing and are now considered as metal and no longer as coins."[38] This understanding applies to the narrative truths of history no less than to the truth of any given object in history, to the "fact" of history as progress no less than to the "fact" of nature as female. A tropological analysis need not leave off where a given trope has disappeared into "common usage," the moment at which Hallyn leaves his project. Indeed, it is historically instructive to examine exactly those elements of "proper" language that have (in Nietzsche's phrase) become fixed, canonical, and binding for their tropological analogues in seemingly unrelated domains. The purpose is, of course, not just to appreciate the tropological nature of these canonical and binding truths, as if we were merely adding water to something dried up and unrecognizable. Some lost tropes deserve to be decanonized, released, their currency melted down because their circulation is too wide, their operative analogues too many. Some versions of a feminized nature fall into this category as a result of their insidious relationships with the subjugation of women and all other people and things identified with nature. Likewise, the image of history as evolutionary progress, deriving the complexities of culture from a primitive, natural past, has had a decided influence on how people account for their superior cultural advancement at the expense of people and objects they associate with the past.

We are not so far removed from the agricultural history of the American West as it might appear. Nothing could be more prosaic than an extension agent's recommendations to plow deep or the fact that *Bromus tectorum* took over the range of native perennial grasses in the intermountain West. Nothing could be more self-consciously well composed than Frederick Jackson Turner's 1893 elegy to the American frontier. And yet whole seminars full of American historians took Turner's beautiful prose as the truth and were subsequently denounced by more literal-minded historians for having fallen for Turner's rhetoric, his "brilliant and moving odes to the glories of the westward movement."[39] The question to ask is not who is more truthful but how the extension agent and Frederick Jackson Turner (or Turner's favorite poet, Rudyard Kipling) might be writing about the same things. Indeed, in what ways is the (perhaps execrable) poetry of the extension agent so compelling that we can only recognize its tropology with difficulty because it sounds so much like something simpler, objective, or truthful? How is this poetry related to Turner's? How do these disparate styles speak with shared assumptions about social relations and patterns of subjugation, about nature and culture, agriculture and government, the past and the present in the time and place under consideration here? What administrative, ethical, and mechanical technics do they inform?

The chapters that follow address separate agricultural objects as they appeared on the western scene from 1862 to 1945—trees, plows, livestock and grass, and weeds—in an effort to describe what might be called a colonial technics inherent in them that rendered their aggregate agriculture a thoroughly colonial operation. Patterns of domination between classes, genders, and races circulated in the unlikely literatures of forestry, botany, weed science, animal husbandry, and range management. Moreover, these patterns of domination each implied a progressive history from the wild to the domesticated, the natural to the cultural. Agriculture was the means by which an object became valuable (as a crop) and was released from its past into the history of improvement. The technics at stake here is at once ideological, material, mechanical, and administrative. It is also poetic, and it has left its legible traces, the means for understanding its elaborate and multifarious organization, in all the literatures of western agriculture.

Trees

> A common assumption underlying Western thought is that things must have had a beginning.—Vine Deloria, Jr., "Circling the Same Old Rock"

As Robert Pogue Harrison notes, forests have a long Western history as originary places: "From the beginning [that is, during the time of the Greeks and Romans] they appeared to our ancestors as archaic, as antecedent to the human world. We gather from mythology that their vast and somber wilderness was there before, like a precondition or matrix of civilization, or that . . . the forests were *first*."[1] Forests are places where everything else came from, places against which all the animosity of "civilization" turned in a simultaneous nostalgia and contempt for its imagined sylvan origin. Forests have terrified children, confused and threatened the righteous. Forests have also built navies and war planes and many other things useful to civilization according to the Western model. Civilization becomes what it is through the process of deforestation.

As "children of a celestial father," cultural descendants of Greece and Rome have cleared trees to reveal the prospect of God; they have recolonized the earth originally colonized by a primeval forest, counterposing the clearings of family dwellings, cities, nations, and empires to the " 'infamous promiscuity,' " lawlessness, darkness, and disorder of the forest from which they emerged.[2] For Rome, the vast forests of Italy, Gaul, Spain, Britain, and the Mediterranean basin "were obstacles—to conquest, hegemony, homogenization," and Roman administrators "found ways either to denude or traverse this latent sylvan mass."[3] Unlike the Greeks (or more specifically the Athenians), who looked to the sea as their "horizon of destiny," the Romans "had tenacious roots in the unprivileged but reliable soil of Latium, where a rustic people emerging from the forests cleared land for cultivation and loved above all the prosaic results of their labor."[4] They forged their destiny as conquerors and colonizers of land, "clearing it for agriculture and leading to irreversible erosion in regions that were once the most fertile in the world," eventually bringing about their own decline. The Athenians turned their forests into fleets that sank "to the bottom of the wine-dark sea," leaving the landscape around Athens barren and treeless but "drenched with that brilliant Hellenic light"[5] that came to identify its civilization as superior to barbarian darkness. The forest, defined by its inscrutability as an obstacle to decisive or virtuous action, is everything that stands in the way of lateral lines of conquest and civilization and vertical lines of enlightenment and grace.

It is in this context that we can begin to make sense of the seemingly senseless deforestation Euro-Americans impressed on the forests they claimed in North America. This deforestation was especially rapid in the Lake States, beginning around 1870, as the woods were cut to build prairie houses, to fuel furnaces, to lay thousands of miles of railroad tracks across the West, and to hold up thousands of miles of telegraph wire to bring American territory under an efficient system of transportation and communication. Having cut much of the woods over and seen the rest cleared and settled as farms or simply burned, lumbermen moved south and then west by the turn of the century, where they began the process anew. Foresters, at first trained in Europe and by the twentieth century trained in the United States, deplored what they called "cut-out-and-get-out" logging. Whether anybody rejoiced in the revelation of the prospect of God after a clear-cut is hard to say. But it is clear that the developing state required this sweep of deforestation to impose its order on a continental mass, the foresters' indignation notwithstanding. Their work came later, as part of the process of fitting the American forests to the needs, demands, and definitions of the United States as a colonial state. The state had to claim them first.

The idea of the "virgin" forest facilitated this claim, conceptually erasing the fact of centuries of aboriginal forest use and manipulation. The aboriginal forest had to be, in Gilles Deleuze and Félix Guattari's term, deterritorialized—that is, its aboriginal territorial features had to be erased ("decoded") and smoothed over—before the state could claim it or recode it as territory of its own.[6] The state in turn functioned as "an instrument of organized violence," to use Murray Bookchin's characterization, not only imposing its territoriality by coercion but also maintaining its territorial claims, institutions, and sciences by utilizing authoritarian instruments of administration over its own subjects.[7] The most rapid and visible means of writing the state onto western forest territory was to cut it all down and use it up. The forest was thereby transformed and a great deal of useful wood went to build war ships, villages, factories, roads. Deforestation left behind cleared land available for agricultural settlement, where the land would support farming.

This process had, of course, begun in the East, when the state's effective reach across the land mass was comparatively limited. But by 1860, the United States had secured by seizure, treaty, or purchase all the land now occupied by the continental United States. A land mass formerly occupied by heterogeneous groups of often migratory people was at least abstractly transformed into property, a preliminary capture of land that allowed the state to supervise and support continental occupation and sedentary settlement by its own subjects. The transformation of aboriginal territory occurred rapidly after 1860, and the deforestation that had trundled through eastern forests in two and a half centuries swept through the forests of the Lake States in about twenty years. Northwestern forests were not far behind, bearing a lumber boom like that of the Lake States shortly after the turn of the century.

It was only when enough trees clearly had been cut in this process—that is, when the "virgin forest" had been decisively deflowered—that the complicated work of reterritorialization began. The state couldn't do without wood any more than it could do without territory. For the purposes of this chapter, then, the story of western trees is understood as a story about the territory of the state and its first order of business in a process of authoritarian, colonial conquest in forested areas. We might think of this conquest as primarily Roman rather than Athenian, land-based rather than sea-based colonization that was perhaps less concerned (in the matter of western forests after 1860 at least) with the question of the visibility of God.

Approaching western forest history as a history of territory addresses a major problem of other ways of thinking about it. First of all, it is tempting to see the lumber barons as unenlightened spoilers of the public domain whom the wise foresters of the Forest Service attempted to educate and restrain; or, failing to exonerate forestry and the Forest Service, it is tempting to see timber companies and foresters as in cahoots against the "wilderness," which would be more properly protected and studied by the Department of the Interior and its specialists than by the Department of Agriculture.

Neither of these points of view fully accounts for the proliferation of the powers of the state over forests, which now include not only the power to unleash private deforestation but also the power to replant forests by design, schedule their harvests, set them aside under the supervision of various bureaus, send civilians and soldiers into them to fight fires, and use them to continue the production of knowledge that will support further measures of control and supervision. In order to loosen the state's hold on American forests, the territoriality of the state (which is always authoritarian) would have to be challenged. Conservation and preservation (even preservation of "wilderness"), as tenets of state forest policy or "private" lumbering, are incorrectly understood as stays to the hand of wholesale forest transformation. In fact, these measures confirm the ambition of the state (however absolutely unrealizable) to control every aspect of the forests within its territory.

Foresters tell a remarkably formulaic story about the development of their discipline and the idea of "forest policy." George Perkins Marsh, whose influential book *Man and Nature* appeared in 1864, noted that the word "sylviculture" had yet to become common in English because this art and science had "been so little pursued in England and America."[8] According to foresters' own accounts of the history of their discipline, forestry came late to England because that country relied for centuries on importing wood rather than growing it at home. Forestry came late to the United States because the forests growing on so much of the continent seemed inexhaustible.[9] Only when people were faced with a timber famine, wherever they lived, were they likely to implement public policies to protect timber or educate specialists to encourage its growth.

For William Greeley, chief of the U.S. Forest Service from 1920 to 1928 and an indefatigable advocate of reforestation, forest policy in general was "an outgrowth of the unremitting pressure of people upon natural resources."[10] The first professionally trained forester to head the small department that became the Forest Service, Bernhard Fernow, understood the development of forestry as a discipline in the same terms:

> Forestry is an art born of necessity, as opposed to arts of convenience and of pleasure. Only when a reduction in the natural supplies of forest products, under the demands of a civilization, necessitates a husbanding of supplies or necessitates the application of art or skill or knowledge in securing a reproduction, or when unfavorable conditions of soil or climate induced by forest destruction make themselves felt does the art of forestry make its appearance. Hence its beginnings occur in different places at different times and its development proceeds at different paces.[11]

Foresters look back on the period before "necessity" made people change their ways as a less mature period, one characterized by the "myth" of inexhaustibility (as opposed to the self-evident realism of scientific management) or by forest uses typical of people less "advanced" in civilization. Shirley Allen revealed the form of contempt reserved for less "mature" attitudes about forestry in his 1950 forestry textbook by comparing American timber consumption with "the Indian who consumed his noon lunch on the way to work in the morning" and "the Negro who scorned an opportunity to earn a quarter because he had a quarter."[12] The development of rational forest uses and policies corresponds with the development of civilization, specifically in the form of the Euro-American nation-state, understood to occur as inevitably as does any process of maturation.

It is only in the "early, thinly-peopled stage of modern nations" that forests are used to provide game, mast, and forage, with the implication that these uses of forests—not readily distinguishable from aboriginal uses—must eventually give way to tillage and pasturage and later to the "indiscriminate cutting of construction wood and fuel," which is the immediate precursor to regulations characteristic of the "modern nation."[13] In the nationalizing push to develop land, "the forest appears an undesirable encumbrance of the soil, and the attitude of the settler is of necessity inimical to the forest: the need for farm and pasturage leads to forest destruction." Fernow, like Greeley, linked the successive phases of forest use and eventual management with "all other economic as well as political developments."[14]

Forest policy, once it is formulated, is therefore "often an expression of national character."[15] Not surprisingly, we see a preoccupation with American exceptionalism in historical accounts of American forestry. As W. N. Sparhawk put it in the 1949 *Yearbook of Agriculture*, which was devoted to trees:

> Several salients stand out in the story of how forestry and the country grew up from a spoiled, wasteful childhood to a rational adulthood. In its broad outline, forestry in the United States is evolving much the same way as it did in Europe, but much faster. Forestry in America has not caught up with the more advanced European countries, but we have come a long way in our brief period as a Nation, and the progress we have made came not from slavishly copying the European pattern; American forestry, as it grows to maturity, tends more and more to become indigenous.[16]

The early "need" for farms and pasturage is taken to be self-evident. It is sometimes seen as the result of population growth (as if this were merely a natural phenomenon),[17] but agricultural development could just as easily be understood as the cause of it. The idea of the "development" of civilization through the process of resource exploitation and the spread first of agriculture and then of industry is so naturalized that foresters simply fail to ask the question of how such a "need" arose. In any case, the stages or phases of development cited by many foresters include a stage of forest exploitation as a natural(ized) sign of national development. You can't have the "modern nation" without the "need" for farms, pastures, fuel, and construction wood. The vehemence with which foresters have deplored the penultimate stage of forest use must be understood in this context of national progress: to be driven to the point of timber famine by the expansion of agricultural settlement and industrial exploitation of the woods is an indication of the progress (technological, cultural, economic, and bureaucratic) of a state engaged in colonizing new territory.

Before foresters could be the harbingers of the happy outcome of that process, a great deal of deterritorialization had to take place. The United States transferred about a billion acres, or two-thirds of the "public domain" (claimed from previous inhabitants),[18] including four-fifths of all of its timberland, to states, individuals, railroads, and other corporations between 1841 and 1891.[19] The relationship between migratory lumbering and colonial nation-building was based in large part on these grants of western lands made by the federal government to subsidize the development and settlement of the nation's interior.

Lumber companies sought large tracts of land recently in the public domain, much of it on property owned by railroads, but they were not above simply stealing timber from land still held by the state or amassing large properties by fraud. It had been illegal to cut timber on public lands since 1831, and every land law carried provisions limiting the use of land acquired by the law, but the laws were simply not enforced in an era in which the federal government itself was encouraging the rapid settlement of the West. The large-scale land acquisitions made by timber companies, supplemented by theft or fraud from the public domain, contributed to the growing nation's occupation of millions of acres of territory in a relatively short period of time. Migratory logging transformed western timber very rapidly, then, into wood products that could be used to build various pieces of the national infrastructure and left land cleared for farming as a by-product wherever the soil would support agricultural occupation.

Through the 1860s, as more people moved toward and beyond the Mississippi River, the process of deforestation moved with them. Deforestation changed somewhat in pace and character when it moved west. Lumbering had always been a commercial operation; it was the first European industry established in America, and the manufacture and sale of wood products had created profits for sovereigns and landowners (and timber shortages for easterners) in trade with Europe and elsewhere since at least the seventeenth century.[20] By 1870, though, the value and efficiency of the industry had increased dramatically, and the availability of rail transportation made it profitable to cut timber in previously remote areas, far from either settlements or river or lake transportation routes. Lumbering had become the most valuable industry in the country, with a gross product in 1870 worth $250 million, sold to build new communities and thousands of miles of railroads and telegraph lines.[21] The introduction of crosscut saws, steam "donkey" engines for skidding logs, and logging railroads to take timber to mills and to other lines for transport to markets all accelerated timber production and transportation. Public domain stumpage (that is, the right to cut timber on public lands) was available for as little as twenty-five cents per thousand feet of timber, which encouraged speculation.[22] Timbered land, infrastructure, and capital were concentrated in the holdings of fewer and fewer corporations. Timber could be liquidated quite rapidly, and what Greeley called the "lumber front of big mills and venture capital" became migratory, decimating the forests of Michigan, Wisconsin, and Minnesota by about 1890. The center of the highest volume of logging then moved again, to the pineries of the South in the 1890s and in another twenty years to the old-growth conifer forests of the Northwest, whose large, high-quality timber remained in demand through World War II.[23]

Deterritorialization is by nature disorderly; it is the undoing of other territory, a way of sweeping territory clean in all directions, unraveling it. Migratory logging was a deterritorializing force unleashed by the state that spread over the timberlands of the West. It was a force that the state encouraged, by outright grants and offering cheap stumpage, specifically to subsidize the construction of transportation and information infrastructure. In these ways, the state always recaptured some benefit to itself from a force over which it had incomplete control, as is clear from the de facto suspension of the land laws. Migratory logging was a moving force that performed the necessary task of removing the forest and providing building materials, but it nevertheless challenged the state's efforts to control migrations, "vanquish nomadism," as Deleuze and Guattari put it, whether in the form of aboriginal nomadism or in the form of the partially necessary migratory nature of mid-nineteenth-century logging.[24]

States are in the business of *settling* territory—in the sense of stilling it, stabilizing it, holding it under systems of sedentary occupation—or at least unsettling territory in controllable ways. Deleuze and Guattari describe this ambiguous condition as a process of "capturing flows," that is, not stopping the movements of "populations, commodities, or commerce, money or capital, etc." (including, in this case, migratory lumbering), but regulating them, directing or redirecting them, like a system of pipes and valves that only directs a flow without controlling what's in a pipe or preventing its occasional escape.[25] With respect to western logging, it was only a matter of time before the state would claim forest territory directly for itself as a distinct overcoding of the blank "public domain," pooling migratory logging in private holdings in order to stop its free flow over the previously uncoded territory. Having corralled private logging and seized forests for itself with the creation of the first forest reserve in 1891, the state could extend its power over forest territory directly without having to resort to the forces of commercial logging that were so difficult to control.

The desire to create forest reserves came at a time when forestry, as a scientific discipline, promised a relatively new form of control over forestland besides the wholesale liquidation of trees. In contrast to what we might call macromanagement of forest property, the sciences of dendrology and silviculture opened the possibility of micromanagement of the trees themselves. Dendrology came first, or that branch of nineteenth-century botany that identified, named, and studied the biology of trees: "The nature of the plant material, its biology, its relation to climate and soil, must be known to secure the largest, most useful, and most valuable crop; that portion of botany which may be segregated as dendrology—the botany of trees in all its ramifications—must form the main basis of the forester's art."[26]

Various collections of information on the biology of American trees were published, one of the earliest comprehensive studies being C. S. Sargent's *Silva of North America*, a set of "bulky volumes" published between 1890 and 1902.[27] Another comprehensive dendrologist, George Sudworth, developed a special emphasis on the forests of the West—an emphasis that was not surprising since these were the forests most recently brought into the national circulation of logging companies and liquidated timber. In addition to his basic *Check List* cataloging American trees, published between 1886 and 1898, Sudworth wrote the monograph, *Forest Trees of the Pacific Slope*, and a series on the conifers of the Rocky Mountain region. Having laid an "excellent foundation," Sudworth died in 1927, after which basic dendrological work like his nearly ceased.[28]

Bernhard Fernow called silviculture "applied dendrology," extending the focus beyond "the general features of the biology of the species" to include virtually every aspect of a tree as a thing that was not merely made of wood but *produced* wood: insofar as the forester is "a producer of material for revenue, he is most emphatically interested in the amount of production and the rate at which this production takes place. . . . It is of importance not only to know the likely progress of the crop, the mathematics of accretion, but also how its progress may be influenced."[29] Silviculture, then, included virtually every factor that could influence the capacity to produce wood—all of the "habits" of trees: "their seed production, rates of growth, reproduction habits, and their reactions to different soils to drought, deep shade, frost, and other climatic factors, to diseases, insects, and fire."[30]

Early work in this area appeared shortly after the turn of the century and tended to be observational in nature rather than describing a comprehensive silviculture. The role of the forester in promoting forest regrowth by using various methods of cutting (including clear-cutting) and then allowing the forest to regrow on its own or replanting it was the focus of other silvicultural work begun about the same time, including Gifford Pinchot's 1905 *Primer of Forestry*.[31] Trees had ceased to be merely material. They had become agricultural objects, with biologies that could be fostered or interrupted to the advantage of the forester. More wood could be obtained by managing the trees' own capacity to produce it instead of simply by finding more trees. The idea of timber as a crop, a forestry commonplace, represents less a stay to the hand of the wholesale transformation of the forest than a proliferation of the power foresters could exercise over the forest to change it in more subtle ways, ultimately for the same purpose: to take its wood.

From the beginning, American forestry was a state science enmeshed in a growing federal interest in claiming forests for the state. The bulk of forest acreage in question lay in the West, where the federal government still held most of the land in the public domain. Early advocates of forest preservation were amateur foresters, but they nevertheless engaged in their avocation as agents of the state.

Franklin B. Hough, for example, was a physician, avocational historian, naturalist, and statistician who became interested in forest preservation as superintendent of the New York State census. He was a member of a commission appointed in 1872 to investigate the possibility of establishing a New York State park in the Adirondacks. He presented a paper to the American Association for the Advancement of Science (AAAS) in 1873 calling for the government to retain its ownership of timberlands in the West and for all states and territories to give some attention to the "cultivation, regulation and encouragement" of forests within their boundaries.[32] Hough's concern for forest conditions was shared by other AAAS members who were professionally trained like himself, although none of them were trained specifically as foresters. These men included Asa Gray, a preeminent botanist; George Emerson, a Massachusetts educator who had published a report of the trees in his home state in 1846; and Eugene Hilgard, a soil scientist who taught at the University of Michigan.[33]

The AAAS persuaded Congress to form a federal commission to study forest destruction and the means to prevent it, and Hough was hired to direct this study in 1876. Only $2,000 was allocated at the outset for this project, and Hough received no further direct funding except for money out of general funds of the USDA until after he presented his reports to Congress. His small Division of Forestry received $5,000 in 1881, $10,000 in 1882, and $10,000 in 1883. Hough published a book and began a journal of forestry while he fulfilled his duties gathering data on forest destruction across the country. In 1881 he traveled to Europe, where professional forestry was well established, especially in France and Germany, in order to bring back useful information. It was only a matter of time before a professional forester took over Hough's duties. He was replaced for political reasons in 1883 by Nathaniel Egleston, a Congregational minister, who was then supplanted in 1886 by Bernhard Fernow, a bona fide forester born in Germany and graduated from the Prussian Forest Academy at Münden.[34]

From that point forward, the Division of Forestry, which became the Forest Service in 1905, was headed by professional foresters, and soon prospective foresters could obtain their professional education in the United States. Bernhard Fernow founded a school of professional forestry at Cornell University in

1898, and the Pinchot family endowed a forestry school at Yale University in 1899, reinforcing the link between federal forestry and the professionalization of forestry as a scientific discipline. Virtually all of the graduates of professional schools of forestry until 1920 went to work for the federal government, and the Forest Service continued to be the largest employer of forestry school graduates through the 1940s.[35] When a group of foresters formed the Society of American Foresters in 1900, they met in the office of Gifford Pinchot at the USDA. The society's members included Henry S. Graves, who became chief of the Forest Service in 1910, and five other Division of Forestry employees.[36]

The Division of Forestry could do little besides gather and attempt to disseminate information without actual forests on which to exercise its growing expertise in forestry. The federal government was comparatively slow in addressing this problem, concentrating instead on encouraging the rapid settlement and development of the West. Finally, in 1891, lobbying by amateur and professional foresters (including Fernow), the lateness of the congressional session, and the length of the General Revision Act (which repealed earlier timber legislation) worked together to permit a section of the act authorizing presidential creation of timber reserves on public lands to pass unnoticed and unopposed by Congress.[37]

The original forest reserves were all located in the West. Later legislation permitted the federal purchase of land for timber reservation, leading to the creation of some national forests in the East. Yellowstone Park was the first reserve established, and the federal government at last had forests of its own to send its foresters into. Between 1870 and 1900, as rival bureaucracies (or individual bureaucrats) attempted to assist either the plunder or the preservation of timber in the West, one phase of land use and occupation was giving way to another. The two were never fundamentally at odds. Migratory, extravagant "depredations" necessarily preceded bureaucratic regulation in the orderly development of the "modern nation." When the forest reserves managed by the General Land Office were transferred to the Forest Service of the USDA in 1905, under the direction of Gifford Pinchot, the federal government's bureaucratic house was set in order as far as forest territory was concerned.

Just as the boundary between state forestry and professional forestry was porous, the boundary between public and private forestland was porous as well. "Private" land, after all, had been distributed by the state from the public domain, no matter how unruly the distribution, and a great deal of timberland reverted to the public domain through tax forfeiture when a timber company abandoned its cutover holdings. Also, private companies routinely cut timber on public lands, both legally and illegally. After the passage of the Weeks Act in 1911, the state could buy back "private" land and return it to the public domain, now designated as national forest. The 1924 Clarke-McNary Act extended the provisions of the 1911 law to include the purchase of cutover lands for the purpose of reforesting them.

The Forest Service, meanwhile, early on had become engaged in the business of transferring its power/knowledge as information to private companies. According to Pinchot, the Forest Service had hoped simply to spread "the gospel of practical Forestry by creating practical examples in the woods" and by sending public foresters into private forests at the request of timberland owners to give advice and assistance.[38] Greeley, however, characterized Pinchot as a fervent conservationist with a "big stick," and even after Pinchot lost his job as chief forester in 1910, he remained on the scene calling for timber regulation.[39] Greeley saw the Forest Service not as a "police officer" but as an educator and facilitator.[40] The Forest Service spent decades trying to bring about the desired final step in mature forest policy—encouraging timber companies themselves to adopt scientific methods of forest management developed by state foresters instead of taking their cut without regard for the future productivity of forestland. The question of whether the federal government should regulate private timber production to bring about this state of affairs more rapidly came up repeatedly from 1924 to 1950, but forest legislation systematically implemented programs of cooperation between timber companies and the Forest Service rather than imposing direct regulation.

Timber companies had always argued that conservation and scientific forestry would come to private forests when it was profitable. Property taxes, the market value of their product, transportation costs, and forest fires all affected the timber industry's profitability. Some wood products companies found reasons to invest in reforestation as early as the 1920s. They started seedling nurseries, planted trees, and employed professional foresters, but clear-cutting was the modus operandi and no other management techniques seem to have been in widespread use through the 1920s.[41] The Weyerhaeuser Company was an exception, investing $1 million in managing 200,000 acres of cutover land in 1924 as an experiment.

Timber companies often forfeited cutover land to local governments for nonpayment of taxes since it was preferable to lose the land once the timber was gone than to wait decades for second growth to make the land profitable again. Weyerhaeuser hired professional foresters to manage cutover forests for sustained yield, and their success on one piece of land led Weyerhaeuser to expand such management to other sites. Other timber companies continued to forfeit their holdings for nonpayment of taxes in the 1930s, but Weyerhaeuser paid most of its taxes, convinced of the profitability of forest management.[42]

Nevertheless, it wasn't until the 1940s, when western old-growth timber became more scarce and more expensive, that the lumber industry in general saw scientific forest management, as opposed to timber liquidation, as both possible and profitable.[43] This was the moment of "necessity," when, according to Greeley's model of the development of modern nations, no great uncut forest remained within easy reach to prevent the move forward to the last stage of development. The public/private boundary with respect to the transfer of forestry information was porous but substantial enough to prevent the wholesale flood of state knowledge/power across it. The relative cheapness of timber before 1940 had also contributed to the lack of forest management, as well as the fact that the state had simply never been powerful enough to apply direct coercion to the forces it had itself let loose.

Although it was desirable for private industry to share the federal foresters' understanding of timber as something to cultivate rather than something to mine, private companies and federal foresters did not have to be in agreement on the value of forest management for all forests to become increasingly enmeshed in projects overseen by various government agencies. This was especially true with respect to those deterritorializing forces that threatened to undo all of the territorial work imposed under the authority of the state.

Fire was preeminent among these forces. Moreover, unlike the flow of migratory logging, which at least built railroad trestles and warships, wildfire could not be captured in any serviceable way. With the introduction of the steam engine, logging itself became a source of this deterritorializing force. Steam-driven logging—with its wood-burning, cinder-throwing engines and its enormous quantities of flammable waste wood, or slash—was a notorious fire hazard. Farmers who cleared timbered land by fire exacerbated the problem. Once the forest had become state or "private" territory, fire was the most feared of those forces that could overrun all territorial designations and destroy property and settlements indiscriminately. It was the only deterritorializing force that rivaled timber liquidators and land clearers in the extent and degree to which forest property could be transformed into a barren wilderness.

As Stephen Pyne's extensive fire history demonstrates, Europeans and aboriginal Americans each inherited traditions of land clearing by fire to establish or nourish agricultural land, create woodland conditions conducive to successful hunting, or thwart enemies. Fire was not in itself an undesirable force as long as it aided some other function and remained under the control of those who set it.

But fire, regardless of its utility, can always "get away." In less sedentary societies, this is not so much of a problem. Indeed, the relatively free play of fire over the space of the prairies complemented the transhumance of Plains societies: both fire and people moved across that space.[44] Fire has always been a weapon of choice against enemies precisely because of its capacity for uncontrollable destruction. Within the institutions of the "modern nation," however, fire is better confined to furnaces, spark plugs, or the precisely directed explosion of powder behind a bullet. In the earliest stages of colonial settlement, fire (like logging) was one force available in the deterritorialization of aboriginal territory. As the state established more and more of its permanent apparatus, however—its cities, highways, railroads, farms—fire lost favor among practically everyone who was not still clearing land.[45]

At the turn of the twentieth century, forests were among those features of the state designated as permanent. For timber companies, fire represented an enormous loss of property because no forests existed further west to move into. Before that time, northeastern farmers had been accustomed to firing the woods to clear fields, and northeastern loggers were familiar with the threat this otherwise desirable fire posed to merchantable timber. When both abandoned the exhausted farms and cut- and burned-over forests of the Northeast for the woods of the Lake States, fire followed them, only this time with the added incendiary threat posed by steam engines and the rapidity with which flammable debris was generated by fast, high-volume logging. "The first railroad entered northern Wisconsin in 1870," writes Pyne, "succeeded a year later by the worst fire disaster in American fire history."[46] The great Peshtigo fire near Green Bay burned 1.2 million acres and killed at least 1,500 people (Native American victims of the white settlers' fires were not counted), annihilating whole towns. Other similar fires swept through millions of acres in Michigan (1871, 1881, 1894), Minnesota (1894, 1910, 1918), and other parts of Wisconsin (1894, 1908).

The fuel for all of these fires had the same source: "Extensive logging created mountains of slash, and natural windfalls added more. Fierce burns left a tangle of downed heavy fuels and thick underbrush. In some areas the debris rose 12 to 15 feet high."[47] Logging slash provided the fuel, and land clearing provided the ignition. Pyne argues that settlers were not properly afraid of the conflagrations that raged around them because fires "were common, holocausts rare." Fires "were a sign of welcome progress" as long as they were used for land clearing.[48] The scale of these fires, however, was unprecedented. They were mass fires comparable to the World War II firebombing of Dresden, Hamburg, and Tokyo. Indeed, the word "firestorm" was first used to describe the 1871 fires in the upper Midwest and returned to use during the bombing of Germany in 1943.[49]

To say that such fires were out of control is a gross understatement. To say, on the other hand, that the horror of the wilderness was reborn in the forest fire reflects the overstated tone of historians' accounts of these fires. Compared with "marauding savages," forest fires had killed more pioneers.[50] These fires were characterized routinely as holocausts and the burned trees and villages as sacrifices, invoking at once the totality of destruction as well as the value of the board feet of timber and the logger/farmer settlements that had been destroyed. Forest fires also invoked images of the sublime. The Peshtigo fire threw out a "roar that was greater than Niagara's"; the 1894 fire near Hinckley, Minnesota, terrified threatened residents by its "far-off roar that sounded like a great waterfall."[51]

Forest fires pushed their historians to use the strongest language at their disposal to describe the fires' fury: they "sounded like thunder, the pummeling of a dozen cataracts, the pounding of heavy freight trains, and 'all the hounds of hell.' "[52] The force of such fires, which burned not only trees, wildlife, villages, and their inhabitants but also the soil itself, threatened to undo all of the layers of property and colonial use under which timbered midwestern land was ordered and received its value. Fire could escape every abstract boundary and render an entire landscape homogeneously barren.

It was not surprising, then, that the public/private distinction would be especially porous with respect to fighting fire as a common enemy. Regulation of burning practices and settler cooperation in fighting fires was by no means new, but by the twentieth century, the tolerance for any fire had decreased, as both settlers and federal bureaucracies understood the period of land clearing to have ended. As the federal government claimed forests of its own beginning in 1891, it was more directly involved in the problem of runaway fires in valuable timberlands. High-volume, highly flammable logging operations moved into the forests of the Northwest around the turn of the century, and forest conflagrations soon followed. Fire was such a common scourge in western forests that residents used it as a justification for cutting as much timber as they wanted since it was likely to burn up anyway. Documenting western resistance to paying for timber rights, John Ise wrote, "One reason why the western men felt that they were entitled to free timber, even for manufacturing purposes, was that forest fires were destroying immense amounts of timber each year anyhow, and there was no reason why this timber should not be used rather than allowed to go up in smoke."[53]

Without fire control, federal forestry and industrial logging were equally impossible. The first handbook on the management of forest reserves in 1901 stated explicitly that "the first duty of forest officers is to protect the forest against fires."[54] Private timber companies took the initiative to organize fire protection for their own holdings, setting up patrols during particularly dangerous seasons and equipping and training fire-fighting crews. Fire protection associations were formed in many western states. Among these associations, the Western Forestry and Conservation Association (WFCA), created in 1909, stands out for its efforts to encourage public and private cooperation with respect to fire protection and to urge its member companies to abide by the principles of scientific forestry. The WFCA's membership included Forest Service employees E. T. Allen, district forester for the Pacific Northwest District, and William Greeley, who later became chief forester.[55]

The Big Burn of 1910 brought all of the components of twentieth-century forestry together in a telling trial by fire of the emerging national bureaucratic regime of forest management. Private and public interests, all levels of government, foresters, labor, and the military joined in the first organized effort to coordinate and integrate authoritarian power to combat the common "enemy," fire.[56]

In 1910 the state of Washington recorded the lowest precipitation in forty years, and the weather was unusually hot. By June fires had flared up in Washington, Idaho, and Montana, many of which had not been contained by firefighters. William Greeley sent temporary workers to protect the Blackfoot and Flathead National Forests from the spread of fire. He later wrote, "We cleaned out Skid Row in Spokane and Butte" to fight fires in 1910, a strategy that was inefficient from a management point of view since these workers weren't considered reliable. They could make a quarter a day fighting fires and were often accused of setting fires themselves to ensure their employment.

The work of better-organized "splendid crews," accustomed to oversight and employment, sent out from logging and mining camps was more effective. After the WFCA called on President William Howard Taft to send federal troops to help, a hundred African American soldiers from the Twenty-fifth Infantry Division arrived on 10 August to ensure the orderly evacuation of towns and to fight fires. In the course of guarding the evacuation trains, they selected men out of the crowd of evacuees to work as firefighters. Wallace, Idaho, the largest of the threatened towns, declared martial law on 21 August to prevent looting. Forest rangers supervising firefighters emphasized obedience and organization. One ranger, Edward Pulaski, became a folk hero, saving his men from oncoming flames by ordering them into an abandoned mine; a firefighter who disobeyed died within minutes. Another supervisor said to his crew, "Obey me and we'll all live."[57] Other crew bosses threatened their men with weapons in order to ensure their tractability.[58] The importance of obeying official authority became a stock component of what Timothy Cochrane identified as the genre of fire stories.[59] (Norman MacLean's account of the 1949 Mann Gulch fire in Montana is in this genre.)[60] Before the weather changed and rain and snow fell on 23 August, fires had burned millions of acres, had cost $800,000 to fight, and had killed at least eighty-five people, seventy-two of them firefighters. Most of the dead were working-class European immigrants, "transients" brought in from western cities to fight the fires.[61]

The fires had been enormously destructive, but the effort mobilized against them brought a coordination of bureaucratic power and authority into the woods that remained intact. Journalist and popular historian Stewart Holbrook described the undertaking in military terms: "[Forest supervisors and] an army of 10,000 rangers, loggers, miners, ranchers, and laborers, plus several companies of soldiers, were to stage the first organized and large-scale battle against forest fires in the United States."[62] What was armylike about this coalition was the fact that it faced an "enemy," a commonplace explicit in the title of Charles

Cowan's book on forest fire, *The Enemy Is Fire*, and evident as well in William Greeley's understanding that "first and last, fire has been the great destroyer of American forests."[63]

Like the real army, the forest "army" was composed of men, a fact that was not altered for decades. The presence of soldiers among the firefighters reinforced the military effect of the forest army. Also like the real army, the forest army was organized and disciplined along lines of bureaucratic authority and obedience. Before 1910, settlers and timber companies either simply fled fires or battled them in a less organized manner. By establishing a hierarchical organization, the Forest Service attempted to guarantee an orderly, planned response to fire, a project in which the military seemed more useful than disorganized, ad hoc transient labor. After 1910, William Greeley proposed that the military establish outposts in the national forests during fire seasons so that its better-organized and more easily controlled labor would be close at hand.[64]

Whatever foresters trained in the subtle arts of dendrology and silviculture might have thought about timber companies' liquidation of western forests, they understood fire to be more destructive. This coincidence of belief led to further cooperation between timber owners and the federal government in an increasingly integrated system of forest "protection" and fire control that extended throughout most of the timbered land of the West.

One of the first results of the 1910 fires was the willingness of Congress to support forest preservation legislation outside of obscure appropriations bills and riders. Congress passed the Weeks Act in 1911, allowing the federal government to purchase land for national forests and granting federal money to states willing to implement cooperative plans to prevent and fight fires. More interestingly, it was fire control, after 1910, that brought previously remote western lands under federal control and supervision, at least as far as fire prevention was concerned. As Pyne put it, "An administrative presence was required: the public demanded signs of 'management,' which it perceived from public words of foresters to mean fire protection; and the canons of professional forestry insisted that fires had to be tamed before the land could be used for the expressed objectives of scientific stewardship, that is, of conservation."[65]

Roads, lookout towers, equipment depots, telephone and radio communication networks, and ranger patrols connected forestlands that were otherwise unsettled and inaccessible. Although the military played an important role when called in to help the Forest Service fight fires, the strategic system of intelligence gathering and supply distribution set up in western national forests did not simply emulate military security. It was an extension of the federal control of land that subsidized the building of railroads, supported settlers' claims to agricultural lands, and sent the military out to vanquish whatever aboriginal populations obstructed the federal reach. After 1910, the vast forests of the Northwest, regardless of the presence of settlers, aborigines, or logging operations, came within that reach under the agency of the Forest Service.

Gaining control over the threat of fire that had crystallized the presence of the state in western forests continued to be a federal priority through the 1940s, but federal bureaus also had other opportunities after 1910 to integrate western forests and state forestry within authoritarian systems of control of both trees and the people working among them for purposes other than fire control. Military involvement in western timber production and forestry, as well as fire control, increased after 1910 as specific projects of the state—including wars against foreign enemies as well as efforts to reduce domestic unemployment and natural resource degradation—demanded the mobilization of both trees and labor. The military presence in western woods after 1910 and military uses of western timber and state foresters together underscore the importance of western forests to the state as a matter of national security.

State involvement in timber production, wood products research, and the containment of labor activism during World War I was an example of the authority imposed by the state through both the Forest Service and the military on western forests and forest workers. The war effort significantly blurred the distinctions between Forest Service and military personnel and civilian and military forest research. When the United States entered the war, the War Department named a committee of lumbermen to coordinate timber production for military purposes and appointed William Greeley as chairman. It was the job of this committee to fill orders for trench lumber, entanglement stakes, road planks, and other wood supplies needed in Europe. Forest Service chief Henry Graves and his successor Greeley were both instrumental in organizing a Forestry Division of the American Service of Supply to coordinate timber production in Europe, and by 1918 the division operated ninety sawmills behind the trenches in France.[66] Greeley was sent overseas as a colonel in the Twentieth Engineers Regiment, managing the production and milling of timber from an office in Tours.[67]

In the United States, timber production accelerated in the West as "both men and natural resources . . . were mobilized"; logging "became more than a mere matter of business—it became a patriotic duty."[68] Civilian use of lumber was curtailed. War production shipped the equivalent of a full year's supply of civilian timber to Europe in the form of weapons and provisions.[69] During World War I, the federal government turned logging over to the army. The Spruce Production Division of the Signal Corps was created in 1917, which produced 180 million board feet of timber, 120 million of it for the Allies. The army's aggressive logging of western forests—"thorough and tragic"—alarmed the Forest Service, which was prevented from supervising logging operations,[70] but conservation was as incompatible with the war effort as it had been with territorial occupation.

The military was interested in more than wood supplies. The war made "numerous and important demands on Forest Service technical knowledge," as Chief Henry Graves noted in his 1918 annual report.[71] The service's Branch of Research was still quite new, established in 1915 to coordinate research at experiment stations in national forests and disseminate their findings to private companies. After 1917, the Forest Service redirected the focus of all of its research on war-related problems, such as preparing special materials for wagons and other vehicles, rifle stocks, airplanes, and shipping containers. The war effort expanded the Forest Service's wood products research capability significantly, including its operations at the Forest Products Laboratory in Madison, Wisconsin, where researchers studied problems of airplane construction, among other things.[72] Military appropriations for wood products research did not end after the war, as the War Department continued to commission studies of wood properties for use in airplane construction. At the same time, war research led to the dissemination of information about new wood technologies to wood-using industries.[73] These technologies included new processes of kiln drying that reduced the amount of time required to season lumber and new methods of construction, properties, and uses of plywood, both technologies used by airplane manufacturers during the war that spread to other industries later.[74]

The accelerated pace of western logging brought about a crisis over the issue of labor rights in the Northwest. Loggers had long been fighting for shorter working hours (the average workday for laborers in the country's most dangerous occupation was ten hours) and better food and living quarters, but without much success. During the war, the Industrial Workers of the World (IWW) organized western loggers as well as miners and railroad workers. A loggers' strike "all but paralyzed" the cutting of Douglas fir for the war.[75] Ship carpenters refused to work with lumber from "ten-hour mills."[76] The tension between Wobblies and the companies erupted in violence. The federal government, engaged in raiding IWW offices and persecuting its members all over the country, sent an army lieutenant in October 1917 to settle the matter. The Spruce Production Division of the Signal Corps, under Lt. Bruce Disque, in addition to taking control of timber production, was charged with the responsibility of breaking the strike and destroying the influence of the Wobblies. Disque also attempted to force the companies to grant an eight-hour workday. The "solution" was the formation of the Loyal Legion of Loggers and Lumbermen ("4L"), a company union representing both loggers and their bosses that eliminated the IWW and all other unions from the picture for the time being. The establishment of an eight-hour workday, higher wages, and improved camp conditions nevertheless constituted a victory for the activism of the IWW, without which improvements would not have been made. The "4L" remained in place until New Deal legislation banned company unions.[77]

The Forest Service and the military collaborated again in the 1930s, this time in an effort to combat domestic enemies. Their authoritarian modus operandi remained the same as their approach to war production and fire fighting, bringing western reforestation, fire control, and unemployed civilian labor under the coordinated supervision of the federal bureaucracy. Unemployment and deforestation both represented serious threats to national stability and security. The Civilian Conservation Corps (CCC) was organized to address both problems at once. The CCC so completely blurred the distinction between the aims of the Forest Service and the aims of the military in managing labor and natural resources that American pacifists considered it controversial, and the project was increasingly drawn into explicit preparations for war through the end of the 1930s.

The CCC was founded in 1933 as a work relief and conservation agency. Reforestation in western national forests was one of its intended goals, but it was also authorized to perform work on private land (as well as in forests in other parts of the country). Several federal bureaus had a hand in its administration: the Department of Labor selected CCC enrollees; the War Department operated its camps much as it would operate an army camp (except for camps on Indian reservations, administered by the Office of Indian Affairs);[78] and the Departments of Interior and Agriculture designated the work projects. The Forest Service was a major component of the USDA's involvement with the CCC, and the CCC's contributions to western forests were substantial. Interestingly, many people, including President Franklin Delano Roosevelt himself, described the CCC in military terms, even apart from its affiliation with the War Department.

In his message to Congress outlining his plan for the CCC, Roosevelt stated, "The overwhelming majority of unemployed Americans, who are now walking the streets and receiving private or public relief would infinitely prefer to work. We can take a vast army of these unemployed out into healthful surroundings."[79] Among other things, the CCC formed "a standing army of forest fire fighters," particularly valuable in the timberlands of the Northwest. Their tree-planting projects represented "mass attacks" against centuries of timber exploitation.[80]

In addition to its contribution to the wars against unemployment, deforestation, and fire, the CCC aided the primary colonizing work of the state, continuing efforts begun after 1910 to make western forest territory more accessible, primarily for the purpose of fire control. Greeley wrote, "In many sections of the Northwest their shovels and bulldozers cut through hazardous areas of logged-over land and second growth and gave all [private timber companies' fire] protective organizations better access to their backcountry."[81] They also built fire-lookout towers and telephone lines in the western timber backcountry. Perhaps more interestingly, the CCC advertised itself vividly to potential enrollees as a agency that put men to work "amid Nature's Grandeur" restoring the nation's once-pristine natural resources; a promotional photo caption noted, "If there's any frontier left—this is it!"[82] Thus, the CCC was conceptually at work in the "West" in addition to its actual work in western forests.

Controversy surrounding the possible military uses of the CCC had followed it since its inception, primarily because of the role of the War Department in administering its camps. Robert Fechner, the first director of the CCC, insisted from the beginning that workers were not to be given military training. He wrote, "The only thing expected from the men is that they will behave themselves."[83] Although this statement revealed the authoritarian nature of the CCC, it was, of course, a dramatic understatement. Many things were expected of the men that were directly related to military mobilization if not the actual firing of weapons. In a 1940 promotional book for the CCC, Director James McEntee advertised the corps on just those grounds. CCC men did not carry rifles or participate in military drills, "but they do receive training which would be of great value to them and the Nation in time of national emergency," completing 50 to 75 percent of the training they would need to become soldiers. Indeed, this training was required by Congress after June 1940, and McEntee spelled out the CCC's responsibilities for national defense. CCC workers were to be trained in cooking, baking, first aid, operation and maintenance of motor vehicles, road and bridge construction and maintenance, photography, signal communication, and "other matters" essential to military preparedness.[84] The militarization of the CCC was opposed all along by those who had called for the removal of the War Department from CCC administration in the 1930s. Critics lost no time in making the comparison between the possible requirement of military training in the CCC and Hitler's compulsory labor camps.[85]

When "national emergency" loomed in 1939 and 1940, the debate took on more urgency. One enrollee commented on reasons to join the CCC: "If a war breaks out we will be the first drafted and it would be well for us to have some knowledge of how to defend ourselves; then, too, our government has done so much for us, why should we not be willing to show appreciation and train some for possible coming events?"[86] This attitude prevailed. McEntee wrote to potential enrollees in 1940, "A man who can run a bulldozer can run a tank."[87] When the United States entered World War II, the CCC was discontinued. Many of the CCC officers and "boys" performed military duty in the war, and some people believed that by giving more than 3 million men military-style experience and job training, the CCC had helped win the war.[88] In the mid-1940s, the uses of forest labor, western timber, and professional forestry expertise again shifted to war production.

Forests were fundamental to mobilization. Wood and wood products from American forests (as well as forests in other nations accessible to federal authorities and commercial interests) supported virtually every aspect of U.S. military preparedness and action in the war. Military contenders on all sides ransacked their forests for supplies and attempted to guard them from enemy attack. In the United States, the military's search for supplies and lumbermen's fears of attack focused substantially on the forests of the West. Northwestern and Alaskan forests supplied much of the timber mobilized for the war, and the coastal forest of Oregon was the target of incendiary attacks in 1943 and 1944. The pulpwood industry of the South was crucial to the American war effort as well, but this industry was primarily based on trees that were sown and harvested relatively young for pulp rather than for lumber. It was the old-growth forests of the Northwest and Alaska that represented America's real forest wealth, supplying large, high-quality trees for the Allies and a valuable target for their enemies.

The remobilization of western forests for war took place through the agency of many federal departments, private timber owners, and munitions manufacturers. The networks formed between industry, timber companies, and the federal government during World War II entrenched and expanded the federal interest in natural resources, particularly for military preparedness, that had begun earlier in the century. The war effort accelerated timber production and created new logging technologies and industries, all of which affected civilian timber production after the war. Civilian and military uses of the woods converged during World War II. Western forests became more deeply embedded than ever in the military-industrial complex, a self-converting system of industrial timber production that could build houses and merchant ships as easily as it could build barracks and aircraft carriers.

World War II began for American forests in 1940 with an order from the War Department for 2 billion board feet of timber, much of which came from the forests of Washington and Oregon.[89] It would be used to house the rapidly growing armed forces and to support many military technologies, from airplane manufacture to gunpowder. Timber companies organized the Lumber and Timber Producers' Defense Committee in August 1940 to "aid the government in procuring the large volume of lumber and other forest products necessary for national defense."[90] The federal demand for timber was so great and the timber companies' response was so energetic that by November 1940, many acres in the Northwest were in danger of being rapidly liquidated. A series of bills were introduced in Congress to allow the federal government to regulate timber cutting according to Forest Service conservation standards and actually buy timberlands on the eastern slope of the Cascade Mountains in Washington and Oregon to prevent destructive logging.[91] The issue of regulation would remain unsettled throughout the war, presumably because the demand for wood was simply not compatible with conservation. "Sustained yield" in the 1940s primarily meant sustaining the rapid cutting of old-growth forests.

The role of the Forest Products Laboratory (FPL) was as important during the 1940s as it had been during World War I, although the technologies under investigation had changed somewhat. The FPL had developed new wood products for use in airplanes, including resin-bonded pulp for propellers and plywood for wing beams and ribs.[92] Wood-based explosives were also among the FPL's products. As ammunition became more deadly and guns more powerful, stronger explosives were needed to fire them. American troops capturing Saint-Mihiel in World War I fired more ammunition in four hours than had been used in the entire Civil War. There was no reason in 1940 to expect new military engagements to demand less explosive power.[93] Wood chemical technology and the resins, glues, solvents, fuels, explosives, and laminates it produced would play a vital part in the new world war for all of the antagonists.

Just as the FPL had been at work for decades, American forest scientists knew that the Germans had been developing highly advanced wood technologies in 1939, when an article in *American Forests* reported, "A painstaking search has been quietly underway [in Germany] for sources of new products and new uses that will help sustain social and industrial life under the economic strife of peace and war."[94] Germany's wood chemists had produced many items valuable in war, from fibers to fuels, creating the world's most advanced wood technology. In 1939 five new research facilities existed in Germany, and American scientists were warned "not to be caught with inadequate knowledge."[95] Forest products researchers were sure of the utility of American wood chemistry research, particularly with respect to military applications; "it is a safe prediction," one researcher wrote a year before the United States entered the war, "that products from wood that are unheard of today will be used in future wars."[96]

As the war progressed in Europe, and the timber of Scandinavia and Russia was no longer available to France and Britain, the Allies' need for wood intensified timber production in the United States in addition to the demands made on forests by American mobilization.[97] The forests of the West were especially valuable. Old-growth Sitka spruce, found only in the northwestern coastal forests of Oregon, Washington, British Columbia, and Alaska, was cut in great quantities primarily to produce aircraft. Ninety percent of the available stands were in Oregon and Washington. The timber from these stands would not be used to build combat planes, as it had been used when these forests were cut during World War I, but to construct training and other planes.[98]

Sitka spruce, Noble fir, and Douglas fir from the Northwest supplied 100 million feet of logs to be used in airplane construction. William Greeley wrote, "It was the cream of the West Coast Forest. Every stick of 'aero' had to be fine-grained (with at least twelve annual rings to the inch) and straight-grained its entire length. That meant a sixteen- or thirty-two-foot piece of lumber whose grain lengthwise did not vary from absolute straightness more than one inch in twenty."[99] Long, hard Douglas fir boards were needed to replace Asian teak for ship decks. Navy minesweepers needed 110-foot logs for keels. Timber of this quality could only be found in the old-growth forests of Washington and Oregon.[100] Lumbermen there were "the forest soldiers who have been preparing the U.S. Army, Navy and Marine Corps for the biggest job of defense ever known," wrote Stewart Holbrook. The timber they cut went to mills that "drone day and night," operating two and three shifts. Holbrook continued in his characteristically heroic tone: "[Lumbermen were in 1942] working in the ghostly half-light of deep forests that were old when Columbus sailed west, the cloud-crashing timber of the Douglas fir belt, timber so fine, so huge and so ancient as to give even the dullest lumberjack pause when he strikes ax into the foot-thick bark."[101]

Demand for Sitka spruce, and domestic timber in general, was so high that foresters were concerned in 1942 that it might run out in Oregon and Washington and began discussing the transportation of spruce from Alaska.[102] Forty million feet of Sitka spruce logs were rafted 900 miles from Alaska to Anacortes, Washington, in Puget Sound in 1943, "at the cost of a king's ransom."[103] Individual loggers, Sitka spruce "gyppos," scoured the Northwest for single trees and contracted with landowners to cut and haul the trees for a share of what they would bring at the mills.[104] As shortages occurred, some foresters thought forestry regiments might again become necessary to scout out timber and organize local labor in other countries.[105]

Some foresters in the 1940s were concerned that the destructive logging of World War I not be repeated. One forester wrote, "The story of what happened in 1917 and 1918 is only too vivid in the memory of many. Trees by the thousands were felled and left to decay on the ground. Areas were cut clean, the few logs with grain suitable for aircraft taken out, the remainder left where they fell to attract fire, insects and disease. Probably not in the history of American logging—and it has many black pages—has such reckless, useless waste of a valuable and limited timber resource been recorded."[106] Earle Clapp, chief of the Forest Service, believed that regulation of private cutting practices—a pressing issue for the forestry industry since 1940—was urgent in 1942,[107] but regulation during the war appears to have been unthinkable.

Timber companies had been called on the carpet early on, in 1941, for profiteering in defense contracts. Earle Clapp had believed at that time that timber companies were liquidating forests and creating rural slums and that regulation of some kind—perhaps through the states—was necessary to protect the three-quarters of the nation's forests (and the loggers' jobs that depended on them) that were in private hands.[108] Not only did private companies continue to operate without regulation, but the Forest Service itself—the guardian of orderly forest management rather than rapid liquidation—was forced to relax its standards. C. M. Granger reported in 1943, "Under pressure of emergency, the Forest Service has modified some of its regular procedures in order to speed up timber output from the national forests."[109] The Alaskan Sitka spruce had been cut from Tongass National Forest by a private company, presumably under these "modified" procedures.[110] Even the national parks were targeted as sources of military timber. The Park Service, resistant to the pressure of the War Department to open up national parks for the war exploitation of their resources, in 1943 finally allowed Sitka spruce to be cut from Queets Corridor, a piece of land that was scheduled to be added to the Olympic National Park in Washington.[111]

It wasn't until 1944, after 119 billion board feet had been consumed, that the War Production Board considered reducing its demands for lumber. The board estimated that 35.5 billion more board feet of timber were needed but had access to only 31 billion; either consumption would have to be cut, or the supply would have to be augmented with Canadian timber.[112] American timber companies and forest equipment manufacturers congratulated themselves throughout the war for protecting America's "forest wealth" and through their advertisements reassured Americans that such wealth would always exist. But the description of scientific ruthlessness American foresters applied to the German liquidation of Polish forests could just as easily have been applied to American wartime logging, especially of irreplaceable old-growth timber.[113] Americans were their own scientifically ruthless enemies in this respect, although they believed the enemy threat was more intentionally destructive.

As American forests became the valuable targets of rapid liquidation, they became potential targets for the enemy as well, a situation that fostered a lively concern—indeed paranoia—regarding their protection from deterritorializing forces. Fire, of course, was the main concern, and in the 1940s, a new emphasis on human incendiarism as a deterritorializing force violating the territory of the state blurred the distinction between Japanese enemies and American "fifth columnists." Americans were warned to beware of "enemies within our borders" who could "sabotage great sections of the country," especially during the summer and fall fire season. *American Forests* editor Ovid Butler wrote, "Fifth column agents, by planting time bombs or dropping them at night from airplanes into heavily forested regions, could during dry seasons of the year start a reign of forest fires that would throw great sections of the country into panic and disrupt military and peace-time activities."[114]

Foresters suspected that Germany's wood researchers had developed a means of fireproofing German forests.[115] Americans had no such technology. Civilian carelessness with fire was simply intolerable under the circumstances. Illustrator James Montgomery Flagg produced a dramatic poster of a forest fire in 1941 featuring a forest-ranger Uncle Sam pointing behind him at the flames, looking accusingly at the viewer, with the caption: "Forest Defense is National Defense."[116] Matches and cigarettes were seen as bombs dropped by American citizens. Butler wrote, "Never in the history of this nation has the prevention of forest fire assumed such critical importance or commanded such patriotic duty."[117] The U.S. Forest Service, with the help of the Pacific Marine Supply Company (no doubt vigorously engaged in the production of naval supplies from west coast timber), ran a full-page advertisement in 1942 that featured a lurid caricature of Hirohito grinning over a lighted match and a picture of a burned forest; the caption read, "Careless Matches Aid the Axis."[118]

Foresters constructed elaborate incendiary scenarios to justify equally elaborate, authoritarian defense strategies. One forester noted that the enemy could strike by air as well as through the agency of fifth columnists and Axis sympathizers. Enemy planes from bases thousands of miles away could "sow fire pills on our forests," perhaps leaving their pilots at large in the woods, having parachuted to relative safety. He recommended that checkpoints be set up where people entering forests would be asked for their name, address, occupation, driver's license, and "other information," thus keeping potential incendiaries "under the sharp eyes of the forest patrolman and law enforcement officers." In the event that incendiary fires erupted, speed and trained manpower would be required to put them out, as well as advanced technology; the United States could develop a "chemical bomb" to suffocate fire. Incendiary fires had occurred in California in 1917 and 1918, he warned, although it wasn't clear who the incendiaries were or what their purpose was (and the weather was dry). But Americans should be prepared in any case to defend themselves against the "insidious undermining" of fifth column or Axis-sympathizing incendiaries.[119] As Pyne put it, "The upshot was that fire control became nearly a paramilitary service of national defense, wildfire tended to be typed as enemy fire, and the Forest Service more than ever became the centerpiece of the national system of fire protection."[120]

While the most paranoid fantasies of forest patrols did not materialize, the authoritarian response to the possibility of fire was unmistakable. In October 1941, fires that swept through 100,000 acres in California were immediately attributed to incendiaries.[121] As the war progressed, information about western fires (incendiary or otherwise) was censored as intelligence potentially valuable to the enemy.[122] Ovid Butler apparently had considered the possibility of censoring reports of even the risk of fire but rejected the idea: "As for maintaining a policy of silence in respect to the sabotage possibilities inherent in forest fire, the editor holds that danger cannot be successfully met by hiding it any more than can crime and disease. Furthermore, it would seem puerile to believe that enemies within or without the country do not already know of these possibilities and are probably well versed in methods of making use of them."[123]

Fires in southern California in November 1943 thought to be caused by enemy incendiaries prompted an FBI investigation that revealed "perhaps unfortunately" that they were instead caused by Americans' own carelessness. Indeed, the large influx of war workers to the San Diego area who were unaccustomed to respecting the "tinder box" conditions of the fire season had led to fires in an area where they had been less frequent as a result of aggressive antifire campaigns. "War, itself, is the saboteur," one writer concluded.[124]

The Japanese had by this time developed balloons that were launched into the jet stream, carrying incendiary bombs across the ocean. The official incendiary bombing campaign began in November 1944,[125] but fires from such bombs had been found and put out in early 1943.[126] A thousand fire balloons were spotted in North America, from the Aleutians to Mexico and as far east as Michigan, causing 285 incidents and killing six Oregonians but doing little damage to forests.[127]

The United States had the tragic last word in incendiary attacks with the use of the atomic bomb. But the issue of fire prevention at home remained salient through the end of the war, when the Advertising Council—responsible for Flagg's poster and other antifire propaganda of the war years—created the image of Smokey Bear to help keep America's forests green, even after the threat of foreign incendiaries, if not fifth columnists, had been stilled.[128] Two of three national fire prevention programs created during the war to protect western forests remain active in the 1990s.[129]

Wartime logging operations had depleted European and American forests and defied Forest Service conservation management policies. But in the frenzy to secure timber and protect forests, many new wood products and logging and fire protection technologies caught the attention of American foresters and the industries that supplied the military. Men in the woods of the Northwest noticed that their work became more mechanized with the war effort. Stuart Holbrook described the new logging operations with his usual vividness. The woods had become noisier than New York City's "El" with the roars of "juggernaut" tractors hauling huge logs with great speed to what they called the Big Blue Ox—the biggest truck in the world, "it is said," which could carry 120 tons of logs at a time. They also used new motor-driven chain saws instead of crosscut saws. A tree felled in two hours by two men with a crosscut saw could be felled in eighteen minutes by one man with a chain saw. "We got panzers right here at camp Six," one logger said, referring to the big tractors used in the camps.[130]

This was not a strained analogy. A man who could run a bulldozer could run a tank after all. They were similar vehicles, and bulldozers were advertised throughout the war as logging equipment employed in the war effort—maintaining America's defenses at home. A Buckeye bulldozer advertisement from 1944 placed that piece of equipment squarely in a military setting, routing diminutive Asian soldiers from their positions under fire from American snipers.[131] The juggernaut, in this case, was surely intended to crush the enemy; in the forests of the Northwest, it was loggers' jobs that could easily fall before this massive deployment of new, heavy multipurpose equipment.

The logging buildup accompanying World War I gave labor organizations their first foothold in forest operations. After World War II, the Congress of Industrial Organizations (CIO) hired a forester to "guide its policy in matters relating to forestry."[132] At the American Forestry Congress in 1946, convened to bring together members of the Forest Service and the American Forestry Association to discuss the condition of the nation's forests after the war, the CIO sent a lawyer to demand "immediate and complete regulation of forest use by the federal government."[133] The labor unions in the 1940s insisted that the prosperity of their constituency depended on sound forestry practices that created "steadily producing forests."[134] Logging remained dangerous, with the worst accident record of any industry in the country in 1949; the CIO as a result demanded the conservation of loggers themselves in addition to the woods they worked in.[135] This vulnerability of northwestern loggers to the threat of timber shortages and the mechanization of their industry, permanently exacerbated by the mobilization for World War II and the equipment introduced into the industry at that time, still informs the demands loggers make on environmentalists, federal agencies, and timber companies.

As the war transformed the technology of logging, it also revolutionized the labor and tools of forest fire fighting. The shortage of male fire-fighting labor, first of all, made the use of women necessary, although their inclusion did not alter the military overtones of fire fighting in any way. During the summer of 1942, the Forest Service employed women for the first time "on the forest fire front."[136] The changes in American forest fire fighting brought the tools, the techniques, and the language of war into the modern fire-fighting repertoire. Fighting fire became a highly trained specialization that was supported by the actual technology of war as well as the conceptualization of firefighters as elite military specialists through constant comparisons between civilian and military uses of potential fire-fighting techniques.

Before 1933, forest fires were fought by thousands of untrained men. The CCC men who replaced this force were equally ineffective, even if sometimes heroic, being "a bit too young." In either case, large crews of untrained men had always "called for a lot of supervision."[137] The new direction in forest fire fighting used small crews of forty well-trained men who followed orders efficiently. Holbrook characterized them as the "storm troops" of fire fighting.[138] They would be supported by other highly trained specialists who parachuted into fire, a method pioneered by the Russians in 1934 and tested in Washington in 1939 and 1940. Fires could be "blitzed from the air" in the remote backcountry of the Northwest; "lightning attack methods, usually associated with modern warfare, are being introduced with great success."[139]

Foresters understood that military equipment would be valuable in fighting fires at home. Military jeeps, tractors, helicopters (quite new at the time), and airplanes could all be useful in getting men to fires. Tanks could be adapted to carry water. Systems of communication developed during wartime could help coordinate fire-fighting efforts in remote regions.[140] The specialization of fire fighting that took place in the context of new military technologies marked the end of fire fighting as both formal and informal work relief and contributed to the proliferation of objects of federal control—now including professional firefighters and their military-derived tools—in western forests.

Forest products research during the war predictably created hundreds of new materials that could be used during peacetime as easily as during wartime. Bonded wood pulp used in the manufacture of propellers could be used in making knife handles; laminated-paper plastic used to make ammunition boxes could be used to make tabletops and truck floors.[141] Plywood used to construct war planes and barracks could be used to make all kinds of boxes and containers, including prefabricated houses. By the end of the war, so much high-grade veneer lumber had been consumed that the plywood industry was "gradually being forced to resort to patching defects and using less desirable species and smaller logs." It was also beginning to use imported logs as substitutes for domestic supplies.[142]

No one considered taking new goods out of production, and the continuing high demand for wood after the war required increased attention to the methods of logging and replanting—of interest to public as well as private foresters—that might sustain that demand. In 1951 William Greeley believed that the "steady expansion in wood uses and technologies" brought by the war "is the strongest ally of forestry in the United States." He wrote, "Not only are we consuming many more forest products, but we are making many different things from the same tree. . . . War-born uses and markets for wood have given American forestry an economic base it never had before."[143]

The search for materials needed in the war, especially rubber, led to an expansion of knowledge about and access to the timber resources of the Tropics of the Western Hemisphere. Haiti began planting commercial rubber for export to the United States after signing a joint agricultural expansion agreement in May 1941, which the Americans termed "democracy in action." Haiti also established banana plantations as part of the agreement. U.S. capital and "technical assistance" made both ventures possible.[144] The need for rubber in the United States increased the amount it imported from Mexico and Paraguay in 1942 and led to the importation of guayule to be planted in California by the USDA under the authorization of the Senate Military Affairs Committee.[145]

Samuel Record and Robert Hess published a massive compendium of information about the trees of the Western Hemisphere in 1943, updating the last such compendium, which was published in 1924. The new edition "indicates the present and possible sources of rubber" as well as sources of resins, oils, tannins, dyes, drugs, and fibers. Of the 22,000 species treated, 19,600 were found in Mexico, Central and South America, and the West Indies.[146] Researchers' desperation in seeking sources of rubber could be seen in their experimentation with leafy spurge (*Euphorbia esula*), a notorious weed resistant to virtually every effort to remove it whose sap might prove useful in making rubber.[147]

Record and Hess's *Timbers of the New World* followed the tradition of works produced by botanist-explorers of the nineteenth century who found and (re)named the world's exploitable floral resources, such as the dendrological work that had launched forestry in the United States. Its identifying pictures, taxonomies, names, and descriptions were based on a collection of specimens gathered by foresters from the Yale University School of Forestry since 1918.[148] To foresters in the 1940s, the Tropics of the Western Hemisphere represented a still-new, underexploited place about which they knew relatively little. Not surprisingly, Record and Hess understood the accessibility of these forests and their products in colonial terms. "Tropical America is approaching an era of great development," they wrote, in which barriers to the "westward tide of civilization" were breaking down.[149]

The fact that exploitation was the purpose of twentieth-century dendrological knowledge gathering was clear in Record and Hess's introduction to their work, and other foresters understood tropical forests in the same terms. In an article about eastern Cuba entitled "Oriente: New Empire of the Americas," one forester wrote, "On this very hemisphere, in this America which global war has welded closer, there are still almost unbelievable opportunities for exploration and development."[150] Another forester noted that American industry would have to develop relationships with indigenous peoples in new areas and that the Hudson's Bay Company's relationship with Native Americans would serve as a good model. He added that Brazil had already done "commendable" work with the indigenous population through "fair treatment" and active missions.[151] Foresters understood that it was the war that made this new phase of colonial exploitation "necessary" and possible.

American forestry had come full circle by 1945. In the forests of the Tropics, foresters and forest industries were beginning the process of territorial control and industrial forest exploitation that had descended on the forests of the Midwest in the 1870s.

From the advent of professional forestry in the United States through World War II, the forces of exploitation and conservative regulation and preservation worked hand in hand to extend an unprecedented array of industrial, military, and agricultural controls over the forests of the West. Although timber barons and conservationists may have appeared to be in conflict over the fate of forests, they merely represented different strategies of control, both of which were supported by the federal government and both of which excluded previous uses of the forest and any uses not explicitly sanctioned by law. Moreover, timber exploitation and conservation together marked the maturity of the modern state. The development of the idea of timber as a crop guaranteed the permanence not of actual forests but of forests as an object of control—something perpetually under the hand of the state or its agents that existed because it was cut and cultivated to order. Whatever agency the forest had to regenerate itself was harnessed to the demands of foresters, to grow timber either for harvest or for "wilderness."

Growing and harvesting timber crops may appear to be more benign than wholesale liquidation, but their more sinister aspect lies in the state's assumption of authority over the conditions of the forest's existence and the maximization of its "efficient" use. Samuel Hays documented the pervasive gospel of efficiency in American conservation and federal resource management and described the privilege and authority scientifically trained experts were to have in making management decisions. Hays argued that this gospel failed to survive the Progressive period because it was politically impossible to sustain centralized authority over all of the resources administered by the federal government.[152]

This was certainly true if only the bluntest instruments of administrative control available to federal bureaucrats and resource management scientists, including foresters, are considered. But the agriculturalization of the western forest represents a more profound exercise of authority than simply delegating power to specialists. Agricultural authority is in part epistemological; it was the silvicultural expertise of foresters that defined the forest as an agricultural place to be managed. Agricultural authority is also predicated on securing territory—against invasion, conflagration, labor activism, indigenous land claims—mainly by force and guarantees the territorial stability needed by both the state and its most powerful agents in order to profit. In the absence of territorial crises, it might seem that the specialists' work was the most significant exercise of control over western forests, whether they worked in public or private offices.

Such ambitiously systematic control of resources, however, cannot take place in a social vacuum. Any threat to the permanent control of the western forests, or any sudden demand for accelerated production, amplified a whole system of authoritarian social relations that already existed—not only between "experts" and everybody else but also between military officers and enlisted men, bureaucrats and "boys" from families on relief, timber owners and loggers, forest rangers and transients.

The western forest provided an opportunity for the United States to express its maturity as a state in several ways. The first was its commitment to claiming and permanently enclosing western forestland, not just using it but institutionalizing it as property. Lumbermen at first did not claim property, just trees; in western forests, the federal government was the first to claim property and set it aside for its own purposes in designated forest reserves. The agriculturalization of the forest, begun by federal foresters and taken up by timber interests wealthy enough to invest in forest property, multiplied the ways in which a forest could be permanently controlled.

Reconceiving the forest as a "crop" had more to do with establishing an increasingly sophisticated system of forest administration than it did with relaxing demands made on the forest. Indeed, it was the permanence of the state (and its endless need for wood) that demanded the permanence of the forest—that is, permanent access to forest resources, theoretically guaranteed by "rational" principles of timber cultivation and regeneration. The bureaucratic and social power of the state was exercised through the complex, coordinated work of fire control, conservation, and timber mobilization for war, all projects in which "private" and federal interests were closely integrated. Where federal and private interests coincided, the institutional power of the state to order social relations and the use of resources was most clear. People were organized by authoritarian institutions, including the military itself from time to time.

Trees were absorbed into these institutions as one more class of objects under careful management—counted as property, guarded against fire and waste, nurtured as a crop. The rapid and widespread liquidation of trees marked the first stage of colonization in many parts of the western forest, but with the cultivation of trees themselves, the state and its agents claimed permanent and increasingly detailed authority over objects within their territory.

Plows

> A technology is not comprehended by its physical properties alone. In use, tools are brought into physical relationships with their users. On the largest view, this relationship and not the tool itself is the determinate historical quality of a technology.
>
> —Marshall Sahlins, *Stone Age Economics*

The plow is as powerful as a cultural artifact as it has been as an implement. It represents agriculture on official state and university seals and on the seal of the U.S. Department of Agriculture; the Minuteman at Concord bears a gun in one hand and holds the handle of a plow in the other. "Surely," wrote Charles Little in 1943, "if ever there were a single perfect symbol for the American ethos, it would be the moldboard plow. The virgin American land was made for this plow; manifest destiny was achieved with it; the wealth of the nation depended on it."[1] The polished steel of thousands of moldboards turned over millions of acres of native sod to replace bluestem and wheatgrass with corn and wheat. Native farmers—women—not already overtaken by warfare or disease were encouraged by force to become "housewives," separated from their productive art and expertise, and at the same time Native American hunters were expected to harness horses, plow fields, and raise cash crops.

The plow is more than simply a piece of technology; it implies a system of domestication of animals and people, an emphasis on commodity rather than food production (and a division of labor by gender that removed women's expertise, though certainly not labor, from the field), an ideology of "improvement," a language of cultivation, culture, and work as opposed to wilderness, nature, and idleness. An entire colonial technics is embodied in the plow. This technics, in the broadest cultural as well as technological terms as it appeared in the American West, is the subject of this chapter.

As in the case of the forest, the plow has borne significant cultural and historical weight in the societies that have used it. The social conditions that facilitated plow agriculture had been evolving for centuries in Europe before colonists drew their moldboards through western soil. Thus, before turning to the use of the plow in the American West, it is important to assess the history of the plow and its place in European agriculture and society in order to understand the agricultural priorities and social stratifications that have accompanied it.

Land clearing in Europe was originally accomplished with hand tools and with ards. An ard was a small wooden plow with a single triangular point that was drawn through the soil by a pair of oxen.[2] In pre-Roman times, these tools were used in soils that were easy to work. The hand tools were developed and used by women, who had historically performed most of the productive labor of gathering wild foods and domesticating and cultivating crops, including tubers as well as grains.[3] Maria Mies argues that reliance on the ard represented a displacement of women's expertise (and their original hand tools) in the fields and that its use by men was accompanied by a dramatic decrease in women's social status.[4] Beginning roughly 3,000 years ago, European farmers, who worked with both hand tools and ards (thus women were probably still involved), began a process of crop specialization that led to a decrease in the cultivation of nongrain food plants and the use of wild plants and an increase in the cultivation of grains. The shift may have become necessary as the soil wore out or the climate worsened, since the grains increasingly under cultivation could be grown in poorer soils and cooler climates than other food plants, according to some archaeologists.[5] By the late Iron Age, European agriculture was dominated by the production of bread wheat and the use of ards.[6]

A process that may have begun in response to an ecological crisis resulted in the wider use of men's tools, the cultivation of grains at the expense of other food plants, and the domestication of both animals and women as agricultural laborers. In places where the "beginning of agriculture" (understood as the beginning of large-scale cereal-grain cultivation) has been studied, archaeological evidence suggests that cereal-grain cultivation involved a more hierarchical social arrangement than other forms of agriculture, like the "gardening" tradition of agriculture practiced by women or forms of hunting and gathering.[7] Grain cultivation, long associated with the foundation of "civilizations" as opposed to more "primitive" social arrangements, facilitated the movement of people (including armies) and fostered trade because grain could be stored and transported. It is possible, then, that the forces promoting cereal-grain production were social as well as ecological since in the end they laid the foundation for agriculture as the cultivation of commodities rather than food, in which women were subordinate to men and landless peasants were dependent on the wages and land allowances of wealthier and more powerful landlords.

Although the ard and later the plow characterized the European commitment to grain production and social hierarchy, the absence of the ard does not necessarily correspond with an egalitarian society. American maize, for example, cultivated without plows primarily as food but also as a trade item, was a crop of relatively egalitarian societies including the Hidatsa, among whom women were the primary food producers. In societies where maize assumed a more commercial character, such as the Inca (in which dry kernels served as currency) and Aztec societies, agriculture was a more hierarchical affair, even in the absence of plows or ards. Men cultivated the land of other men, and various strata of men owed tribute to their superiors, paid in maize.[8] In Europe, the ard and other plows simply facilitated an agricultural emphasis on commercial crops rather than food.

Medieval social and technological developments continued along the same lines, always underscoring the primacy of male tools in agricultural work, the enforcement of social hierarchy through economic dependence and violence, and the expansion of grain production for commercial trade.[9] Charlemagne's eighth-century army marked a watershed in European military history, protected by iron armor and brandishing iron weapons.[10] The dependence of the emerging technology of warfare on large-scale iron metallurgy meant that those who were powerful enough to develop mining and metallurgical infrastructure had superior weapons at their disposal to create and enforce social relations to their advantage. As warriors consolidated their territory in the Middle Ages, agriculture intensified. Small plots tilled by individuals with wooden ards gave way to bigger fields of land not cultivated before tilled by large teams of oxen pulling heavier moldboard plows with iron shares.

Ards opened up but did not turn over the soil. Moldboard plows, used by Romans and later by Europeans, had two standard parts: a share that cut into the soil horizontally and lifted it and a wooden flange, or moldboard, that turned the "mold" over. Some also had a coulter—or knife—that cut sod and clumps of soil into vertical strips. These plows could till soil too heavy for ards and could till lighter soil faster than ards since the fields did not have to be cross-plowed to thoroughly break up the soil. Moldboard plows also required greater animal traction to pull them—more animals than were likely to be owned by one peasant family—and greater coordination of peasant labor as well. The moldboard plow was primarily a northern phenomenon. In southern Europe, the soil and climate rendered the ard a "reasonably effective" implement well into the twentieth century. The expansion of grain production in the north, where the soil was heavier and the climate more humid, demanded a different implement.[11]

The resulting agricultural "efficiency" and bounty (in grain) reflected not so much the enlightenment of formerly "backward" peasants charging ahead with technological innovation but the increasing power of warriors and their agents to shape rural society. Territory claimed by peasant use, or even managed collectively by peasants, was reterritorialized by feudal "lords," or bread-masters, a term based on their practice of distributing bread to formerly independent subjects.[12] Warriors organized labor and land use in order to extract rents and surpluses for their own benefit. The old agriculture had provided well enough for an independent population, scattered thinly over the countryside, but it could not support more concentrated and dependent populations or the extraction of rents.

The old agriculture also did not guarantee the submission of those who actually worked the land, many of whom became slaves under the new system. The ironworks that made plowshares for the lord also made his weapons, each of which protected and enriched his domain. Except where people had been able to resist the consolidation of manorial power, ordinary peasants would not regain anything approaching their post-Roman autonomy until the fourteenth-century plague dramatically increased their bargaining power with the landowners. Agriculture intensified again in the sixteenth century, as landowners enclosed property to raise cash crops, dispossessed peasants outright, and proletarianized all peasants except those forced to find a living at their peril in cities or on the roads.[13]

Interestingly, each of these bursts of colonization took place alongside the rise to power of state-minded, bureaucratically sophisticated conquerors: Charlemagne and William of Normandy were both great codifiers of law and legendary warriors, and the Renaissance ascent of the bourgeoisie institutionalized private property and the political power it conferred on individuals. Of equal interest is the fact that better plows and greater efficiency translated into the satisfaction of commercial demands, not a higher standard of living for peasants.

The emphasis on grain culture guaranteed an income for landowners, but for peasants it meant an almost unrelieved diet of grain: "Bread, more bread, and still more bread and gruels."[14] Peasants in other parts of the world whose land has been placed under intensive cultivation of something inedible, like cotton or coffee, have been even less fortunate. Historian Fernand Braudel wrote, "Corn's [that is, wheat's] unpardonable fault was its low yield; it did not provide for its people adequately."[15] Indeed, "All the bread crops added together never created abundance; Western man had to adapt himself to chronic scarcities."[16] Cultural and political theorist Marshall Sahlins observed that the same tendency culminated in contemporary famines: "*This* is the era of hunger unprecedented. Now, in the time of the greatest technological power, is starvation an institution. Reverse another venerable formula: the amount of hunger increases relatively and absolutely with the evolution of culture."[17] Medieval people adapted to agriculturally related scarcity in part by eating crops that fed livestock as well, including peas, beans, lentils, and vetches.[18]

Western man also adapted by relying on Western woman, as always, to provide staple foods apart from grains. Chronic food shortages might be expected in societies relying on an agriculture increasingly ordered by commercial grain production rather than food production as such. Supplemental vegetables were raised on land not devoted to agriculture, relatively small parcels near their houses where women still controlled food production. There women grew cole crops (like cabbages) and onions of various kinds as well as other items needed by their families, including herbs, flax, and hemp.[19] The small scale of this "gardening" must be understood within the context of the exclusion of food production from the fields.

Medieval and Renaissance agriculture institutionalized the primacy of commodity production over food production in Europe, the privilege of landowners, and the marginalization of women in agriculture. Agriculture in the United States took its essential character and purpose from all of these institutions. The techniques and machines developed by Jethro Tull in the eighteenth century completed the basic technical repertoire of European and American agriculture. His influence, which includes the invention of a grain drill, the convention of planting in long, straight rows, the disinclination to rotate crops, and the replacement of hand hoeing with the use of a cultivator drawn by a horse, still shapes the features of "modern" agriculture, although the drills are bigger and the horses have been replaced by tractors. Not surprisingly, Tull based his method of "horse-hoeing husbandry" on one of the oldest agricultural industries in Europe and the classical world: viticulture. Grape vines were planted in rows and cultivated repeatedly with plows rather than hand tools, and they remained where they were planted until they died, without giving way to other crops. His genius was not just in rationalizing grain production but in suggesting that one commercial industry conform to what he believed were the superior cultural practices of another.

A true believer in the privileges of landownership and commercial priorities, Tull was a virulent antirepublican. He detested the practice of leasing property; he also detested servants. He wrote that it might be a good idea "could gentlemen contrive automata to do the business appertaining to tillage without hands [servants], at the price that is reasonable to be given servants and labourers for the same: not that there is any want of hands to receive our money, to take away our goods, and to beat us; but such are wanting as will work faithfully at reasonable wages."[20] Tull clearly understood the direction modern agriculture would take in later centuries.

Eastern and immigrant settlers brought all of these tools, agricultural priorities, and social conventions across the Mississippi River as they sought more and better farmland. Changes in plow technology up to that time had all facilitated greater efficiency of labor but had not altered either the basic structure of the plow or its purpose in cultivating commercial crops.[21] Durable iron plows with steel shares were available after 1807, but they did not work well on the eastern prairies because the rich soil stuck to the moldboards and would not turn over. John Deere secured his fame by developing a polished iron plow in 1837, called the "singing plow," which sliced cleanly through prairie soil. He began using cast steel in the mid-1850s, and other manufacturers did the same in the 1860s. Further improvements were made in the shaping and tempering of both iron and steel plows, and by the time of the Civil War, these were the plows that farmers brought west. Each improvement made the new plows easier to pull and more suitable for prairie soil.

When settlers moved west, including African Americans liberated from plantations where they had primarily used hoes in the fields,[22] they took moldboard plows with them. Settlers' general aims were certainly varied. African Americans fleeing persecution after emancipation hoped to find a better life in the West, and for many of them, even marginal farming offered a degree of autonomy preferable to outright subjugation and pervasive racist violence, which increased in the decades after the Civil War. Hundreds of thousands of African American farmers went to Oklahoma, Texas, California, and Kansas in the 1870s and 1880s. Many European emigrants were trying to escape religious or political persecution or economies that produced endemic poverty, and the uncertain future of emigration to the West seemed better than certain misery where they were. Native American families who began farming with European methods could, with tools of their own, titles to land, and markets for their produce, prevent the most crushing impoverishment—unless they were summarily dispossessed, as, of course, the Choctaw, Chickasaw, Cherokee, Seminole, and Creek were.[23] But all of these people, in addition to those whose primary stake in western agriculture was financial gain, were participating in agriculture on the scale—and with the tools—of commercial production, as part of a society that privileged men's labor over women's, property over labor, and commodities (especially grains) over food.

The 160-acre homestead, institutionalized by the Homestead Act in 1862, was an expression of these technical, social, and economic conditions. Together with the plow technology necessary for a farmer using European methods to cultivate even a quarter of it, the size of the homestead indicates at least two things: the determination on the part of the federal government to recode a "wild" landscape as quickly as possible by creating vast domesticated fields and the commercial nature of western farming. When producing food, a household can live on the grains and food plants cultivated on about one to five acres, depending on the quality of the soil and the skill of the farmer. This must have been as true of Europeans in the remote past as it was for Pima and Papago farmers in the American Southwest and the Hidatsa on the upper Missouri River.[24]

The agriculture that came west with European settlement was by contrast a great devourer of farmland. Each 160-acre homestead on good land could have provided food for thirty households or more if every acre were under cultivation. If only a quarter of that acreage produced food, a homestead might still support eight households. Truly, the homestead plowman had become his own lord and tenant, the breadwinner, taking the produce of land that could otherwise have been divided exclusively for himself and his family. But, of course, any smaller scale of land division was unthinkable, not because all farmers had a conscious or vicious drive to dispossess Native Americans as quickly as possible—although no doubt such a consciousness existed, and violent dispossession befell innumerable people—but because it took 160 acres at least, and often more than that, to support only one family in a society and an agriculture based on the exchange of commodities for cash.[25] Farmers expected to make money on their property, especially since most western land acquired after 1862 was actually purchased rather than granted. And some landowners bought into agricultural production as a financial investment, amassing small empires of their own by the square mile.

As Patricia Nelson Limerick and others have pointed out, the Homestead Act led to exactly this type of consolidation of property in the West.[26] On a 160-acre farm, the men and boys of one family could cultivate their productive land with a steel plow and a team of horses. On bigger farms, the object of farming shifted away from a family's living to the accumulation of great fortunes in commodity production. In this case, the landowners did not farm themselves but hired hands and tenants, relying on a rural, landless proletariat. Implement manufacturers made plows that had multiple moldboards—gang plows—which were operated by gangs of plowmen.

Bonanza wheat farming in Minnesota and the Dakotas in the 1870s was a good example of this type of agriculture.[27] The Northern Pacific Railroad facilitated the project, selling its extensive land grants in the Red River valley and eventually persuading landowners to cultivate them in wheat. A single "demonstration farm" begun by the railroad covered twenty-one square miles, or over 13,000 acres. The first manager of this operation was Oliver Dalrymple, who at the time owned the largest wheat farm in the world in Minnesota's Washington County, encompassing 2,600 acres, or about four square miles. Dalrymple directed planting for the railroad on 1,280 acres in 1876. His first crop was a huge success. Three years later, the demonstration farm sprawled over seventy-eight square miles, two-fifths of it in production. Along the Northern Pacific Railroad line west of Fargo the farm spread out on both sides of the railroad for about twenty miles as far as the eye could see. Other landowners developed their property in great chunks, eight of the largest covering over 300 square miles altogether. Each one might employ 300 men and use 500 horses and mules as well as hundreds of plows and harrows.

Red River bonanza wheat farming did not take place on a primitive frontier. Western farming in general represented the state of the art in land breaking. Although all plows could "break" the soil—that is, break it up and prepare it for cultivation—the breaking plow as opposed to simply the plow became an implement in its own right with the expansion of agriculture onto the Plains. *The Implement Blue Book*, a directory of manufacturers, listed sixty-seven different brands of breaking plows among the hundreds of plows advertised in 1906. Whereas most of the brand names of plows reflected the cities or regions of their manufacturers—such as Manitoba or Moline—the names of breaking plows reflected their use on the western Plains: Cherokee, Jack Rabbit, Prairie Chief, Prairie Dog, Prairie Gem, Prairie King, Prairie Queen, Western Bonanza, Western King, and Western Queen were among the western-style names given to prairie-breaking plows at the beginning of the twentieth century.[28] These plows were manufactured in great numbers for farmers whose agricultural goal was the production of cereal crops. The level surface of the Plains and the absence of trees presented a landscape "ready for the plow,"[29] with no naturally occurring limitations to field size. A farmer on the Plains could use more land for crops than a farmer operating with scattered thirty- or forty-acre tracts. The need for efficiency in production in order to handle large fields profitably created a demand for innovations in all farm equipment. Plains farms became the biggest market for riding plows, multiple-bottom plows, new harvesting machines, and other equipment.[30]

The flood of grain from the Red River valley supported railroads and other conveyances that took the wheat east as far as Buffalo and out into the Saint Lawrence River bound for Europe. It stimulated the flour-milling industry and transformed Minneapolis into "Mill City," the world's greatest miller of wheat. It enriched the implement manufacturers, who sold their products by the thousands and thus could afford to develop bigger and more labor-efficient machinery to plow and harvest crops.[31] An entire infrastructure was in place by the 1890s to support such large-scale farming after the bonanza died out—which it was sure to do. The surplus of wheat drove the price of land up, and the value of wheat down, until many of the big landowners divided and sold their holdings in the 1890s. There had always been more land under cultivation by "small-scale" farmers in the Red River valley than land under the control of the grain barons, but it was the consolidated capital of the latter that ensured the trend toward bigger equipment, cheaper wheat, and larger farms that farmers have been forced to contend with ever since.

The western bonanza farm is an example not of colonization gone wrong but of colonization in its most literal manifestation: enclosing property for the cultivation of commodities. The scale of these operations is bewildering if we think of their produce as food. They were merely a means of industrial production, however, like any factory; they had little to do with subsistence except through the wages they paid to their hands. Western wheat farms merely underscored the trend in agriculture toward large-scale grain production, which had been expanding and intensifying in Europe for nearly two millennia.

Along with the Homestead Act in 1862 came the Morrill Land Grant to fund state agricultural colleges and the establishment of the U.S. Department of Agriculture. The 1887 Hatch Act provided funding to establish agricultural experiment stations in every state. The 1890 Land Grant Act belatedly created agricultural colleges for African American students. Agricultural specialists now had institutions within which they could systematically work to improve the efficiency of tools, techniques, and crops of American agriculture.[32]

The many bulletins and annual reports of the new department and experiment stations also gave agricultural specialists an unprecedented platform to prescribe the values and purpose of American agriculture and celebrate the progress of American civilization. In the first annual report of the USDA, Commissioner Isaac Newton wrote to President Abraham Lincoln, "Agriculture is the grand element of our progress in wealth, stability, and power." He understood agriculture as "a great civilizer in the world's progress" from the "rudest stage" of hunting, through pastoralism and nomadism, to the "spirit of adventure" that resulted in "planting new empires in the wilderness." For Newton, progress was measured by the efficiency of labor made possible by new plows and lighter tools and by the fact that wheat production had increased 70 percent and corn production over 40 percent between 1849 and 1859. To continue this progress, he wrote, farmers needed greater knowledge of agriculture as an art as well as a science.[33] This theory of social and agricultural evolution had a tremendous impact on the treatment of Native Americans in the West by the USDA and the Bureau of Indian Affairs, on patterns of western settlement, and on the direction USDA research took by the end of the nineteenth century.

The federal government saw scientific, increasingly mechanized agriculture as such a positive good that it believed every farmer should practice it, and western Native Americans were among those actively and systematically compelled to change over to new methods. The United States employed similar tactics in other places it encountered "backward" farmers, including the Philippines at the turn of the century. Many eastern Native Americans had been converted to agriculture before the 1850s; the Cherokee, Choctaw, Chickasaw, Seminole, and Creek people became known as the "Five Civilized Tribes" as a result of their early adoption of European-style farming and the accoutrements of agricultural life, including in some cases the use of slaves. Interestingly, although these southern people adopted European-style agriculture, some of them were reluctant to use the plow until the nineteenth century, when a fairly large white market for their crops emerged. Before that, Cherokee and Creek farmers believed the plow would only allow a small number of farmers to produce crops with yields greater than they could possibly sell, forcing farmers with hoes and digging sticks into landless starvation.[34] In the decades following the Civil War, when federal troops accelerated campaigns against Native American resistance throughout the West and forced increasingly large numbers of people onto reservations, the agriculturalization of western Native Americans began in earnest.

The 1887 Dawes General Allotment Act demanded that reservations—which already restricted Native American economies and movement—be divided into individual parcels of land, allegedly to encourage private initiative and to force Native Americans to work their own land as farmers. Many of the allotments comprised, like federal homesteads, 160 acres. No provisions were made for future tribal members. The federal government established a blood quantum to determine who was "really" a Native American, and only those who were one-half or more Native American by birth were to receive allotments. Although many individuals had significant social and linguistic as well as genealogical ties to tribes, without the requisite genetic credentials they were ineligible for allotments, and the "surplus" land on reservations was sold by the federal government and opened to white settlement. Allotments were originally to be held in trust, free from taxes and protected from sale, for twenty-five years, although this provision was removed in later legislation that allowed individuals judged "competent" to sell their land.

In those reservations subjected to allotment, the process resulted in the rapid dispossession of what meager territory the tribes had secured and was characterized by deception and fraud. Robbing Native Americans of even modest community land bases, allotment also often robbed them of a chance to determine their own relationship to a commercial economy. They had to either work their land and make what living they could or sell it. Many were left with the poorest land after allotment, while the agriculturally valuable land (and often the water necessary to irrigate in the West) ended up in the hands of white farmers.

The agriculturalization of Native Americans was, as Douglas Hurt makes clear, largely a failure.[35] Allotment only amplified other problems inherent in the process. In 1860 the most agriculturally successful groups in the West were those who had been plow farmers before removal, having obtained plows from Spanish missionaries. Many more relocated Native Americans, even those who had always been farmers (without plows), had to adjust to soils and climates entirely unlike those they knew best. And Plains people who had not farmed in many years, if at all, were newly confined to reservations where they were cut off from their usual economic activity and forced into dependence on government aid. Some Plains societies characterized primarily by hunting and gathering rather than by farming nevertheless had at one time been horticulturists before adopting the horse and emphasizing the hunt.[36] As historian Preston Holder notes, Plains historians assume that this shift was immediate and widespread among Plains horticulturists. Holder himself argues that some Plains societies, such as the Osage, had always emphasized hunting and gathering (although they traded with their horticultural neighbors) and that apparently it was among these people that the horse became most important. The horticultural Plains societies had by no means given up their villages for the hunt exclusively, although they had ample access to horses and in some cases served as horse dealers to other Plains people. The status of women tended to be higher among groups like the Mandan and Hidatsa who were committed horticulturists than it was among other groups, many of them Siouan-speaking, who were more committed to hunting.[37]

In any case, when Native Americans were approached with the demand to take up Euro-American agriculture, neither historically remote horticultural experience nor women's traditional expertise were of much use in an agriculture in which the men were farmers and the crops and the land were unfamiliar. Moreover, reservations were often located on poor land that would have resisted the efforts of even the most experienced European-style farmers. Help from the federal government was unreliable at best. The Cheyenne and Arapaho were removed to a reservation in Indian territory after 1867 and promised money to buy seeds and tools. They got less than a tenth of the amount promised and were confined to an area so dry that no experienced plow farmer could have made a living there with the technology available in the 1870s. Some government officials decided that stockraising in such places was a better idea by 1872. After all, the Acoma, Laguna, Zuni, and Hopi Pueblos, as well as the Navajo, had all taken up stockraising after contact with the Spanish. The government tried to encourage stockraising, especially among former hunters and gatherers from the Plains such as the Apache, Kiowa, and Comanche, in the 1880s. But often the breeding stock arrived late and was slaughtered to feed starving people before a reproducing herd could be built up.[38]

The Pima, on the other hand, had always farmed and had been influenced by Spanish tools and crops. They became successful grain farmers very early, with ready markets in Spanish and later Anglo settlements. By 1770, wheat was as important a crop as maize to the Pima, planted in fields "so large, that, standing in the middle of them, one cannot see the ends, because of their great length," wrote a Spanish observer at the time. "They are very wide too," he continued, "embracing the whole width of the valley on both sides."[39] The Pima managed to produce valuable commercial quantities of wheat without using the plow. Opening the soil with sticks, they planted wheat, like maize, in hills or clumps.[40] In the mid-nineteenth century, with the acceleration of white settlement and mining and trading activity in the West, the Pima rapidly expanded their wheat production, again with technology that was by that time "outdated." Anglos had abandoned ards a millennia before, but the Pima began using an ardlike plow regularly around 1850 and with it furnished hundreds of thousands of pounds of wheat to the Overland Mail Line, the federal government, private teamsters, and other companies throughout the 1850s. They did not use steel or iron plows regularly until the 1880s.[41]

Rather than seeing the Pima as technologically disadvantaged, we might understand their nineteenth-century use of the ard, and the expansion of wheat production that accompanied it, as an index of the scale of the southwestern wheat market and the extent of the Pima interest in supplying it. They already had a long history of adjusting their own relationship with trade, and they did not feel compelled to maximize wheat production for its own sake or for the payment of tributes or rents. The Pimas' only agricultural difficulty began late in the nineteenth century when white farmers above them on the Gila River diverted irrigation water into their own fields.[42]

By the end of the nineteenth century, agricultural education had shifted from the belated and insufficient grants of tools and haphazard instruction of agents to the compulsory education of Native American children, many of whom were institutionalized in boarding schools like the infamous Carlisle Indian School, founded in 1879 in Pennsylvania. These institutions were supposed to teach Native American children how to work in order to develop "the native 'character,' " but increasingly boarding school work served merely to subsidize the institutions themselves. Children, many of them taken from their parents on western reservations, were relocated and forced to work in order to eat. Girls' needlework and the agricultural produce of school farms worked by boys were sold to support the school. These children became, as Jorge Noriega put it, "functional outcasts in their own society," trained mainly to become agricultural and domestic labor in Anglo-American society, forbidden to practice their traditional work or speak their own languages. Carlisle was so "successful" from the point of view of the federal government and Christian missionaries that it served as a model for many subsequent institutions.

The federal government began using former frontier army posts as Carlisle-type boarding schools after 1883, establishing three—in the Dakota Territory, Idaho, and Minnesota—by 1885. For some legislators, boarding schools remained the preferred method of subjugating Native Americans, by indoctrinating their children, well into the 1940s.[43] But even if boarding school boys did the work on the school farms, the methods they learned were unsuitable in many cases for farming the arid lands to which they would return.[44] The additional irony of teaching Native American boys and men to farm at the end of the nineteenth century was, of course, that the population as a whole was becoming more urban, not more agricultural. Prices for farm produce had been falling since the end of the Civil War. Low prices and farm mechanization on large farms could not support a growing number of farm owners or more farm laborers.[45]

As various federal agencies dispossessed, removed, abducted, and indoctrinated Native Americans, hundreds of millions of acres of the American West were secured for colonization. One enormous problem remained. The "Great American Desert," a dry grassland that stretched from the well-watered eastern prairies to the Rocky Mountains, had been considered inarable since the beginning of the nineteenth century.[46] This was a great disappointment considering that many believed, as had Benjamin Franklin, "the Great Business of the Continent" was agriculture.[47] Plow agriculture made superior use of American land. As settlers moved out toward the hundredth meridian, land that was unsuited for crops might be grazed, but this was a concession to climate to be avoided if possible. Kansans and Nebraskans, for example, did not enthusiastically embrace the possibility that land west of the hundredth meridian would be reserved for grazing homesteads.[48] Where native plants persisted, no advancement of civilization could occur. Wild plants, berries, and roots could "sustain only a roving and more or less savage population, few in numbers and failing to progress from century to century," wrote two agricultural specialists in the early twentieth century.[49]

To use native range as pasture was considered an improvement over leaving the land unused (that is, unplowed), but early evaluations of the Plains expressed some concern about the cultural implications of an agriculture based on grazing. People feared that the "interior basin . . . would be overrun by a strange mixture of half-civilized, pastoral nomadic tribes."[50] The association of pastoralism with savagery helps explain federal agencies' reluctance to encourage Plains tribes to take up livestock grazing. Barbed wire and the industrialization of meat transport and processing would radically depastoralize western stockraising at the end of the nineteenth century. But before that, the advancement of American civilization seemed to be at stake in the effort to promote profitable agricultural expansion beyond the hundredth meridian. Consequently, rather than consign the arid West to bands of nomads—Native American or otherwise—the "imaginative conquest of the desert accordingly took the form of a proliferation of notions about an increase of rainfall on the plains," as Henry Nash Smith put it.[51] Charles Dana Wilber, town builder and amateur scientist, argued in 1881 that rain would follow the plow.[52]

Rain did not follow the plow. But agriculture without plows, or agriculture that focused on grazing rather than cropping land, threatened the definition of agri/culture itself. Wilber's logic was in a way perfectly sound. The plow was the identifying instrument of agriculture, which transformed unimproved land into productive farms. If a lack of rain stood in the way of this transformation, then the plow itself would bring the rain. This logic has never been fundamentally challenged by developments in scientific approaches to farming the dry West. Drought-resistant grains, "dry land farming" techniques, and irrigation all allowed the cleared fields to become productive in a recognizable way. The plow brought the rain if only because, drought or no drought, farmers brought the plow and found ways to conjure crops with creek and river water or what moisture could be deliberately trapped in the soil.

The USDA's Division of Botany began systematic work to bring seeds from other countries specifically to address the problem of western aridity beginning in 1897. Plant "explorers," as the commissioner of agriculture called them, brought drought-resistant grains and forage plants from Africa, Asia, Australia, Central and South America, and the Pacific Islands.[53] Many of these plants, especially varieties of alfalfa, clover, bromegrass, and wheatgrass, would play major parts in stabilizing the livestock industry and revegetating the range in the West after the 1880s. Drought-resistant wheat, primarily from Russia, became the most important crop raised without irrigation in the West by 1910. A great deal of Red Fife spring wheat, originally from Russia, was grown on the northern Great Plains in the early twentieth century and brought the best prices of any wheat on the world market.[54] Winter wheats, some of which were from Russia as well, were very well adapted to the central and southern Plains. When Russian Mennonite immigrants settled in North Dakota in the 1870s, they brought winter wheat with them (as well as some Russian weeds, as we will see in chapter 4). Their wheat became cultivated by a wider agricultural community in the twentieth century.[55] Other wheats from Russia, particularly soft winter wheats, were imported for use in California, Oregon, Washington, and northern Idaho.[56]

Although the Pima had planted wheat for many years without plows, western dry farming with drought-resistant wheats was predicated on plowing. The crops and the tillage went hand in hand and constituted the basic elements of dry farming, which became what Mary Hargreaves called a "movement" to settle the dry West. As she pointed out, dry farming in the sense of raising grain without irrigation was not new on the Plains in the twentieth century. What was new was a dry-farming system.[57]

Wheat farmer Hardy Campbell had begun experimenting with new tillage techniques after five crop failures in the Dakota Territory and had published the influential *Soil Culture Manual* in 1902. He was an unflagging propagandist for his method, publishing a journal, *Campbell's Scientific Farmer*, and several editions of his manual. He commended the work of the USDA in finding new crops but remarked that "it will not do to place great dependence on the finding of plants that will grow in the deserts without application of special methods of cultivation." His work received a great deal of attention in other journals, including *The Nation*, which featured headlines in 1906 that proclaimed, "Dry Farming—The Hope of the West" and "How the Great American Desert Is Being Reclaimed without Irrigation." He became a consultant for railroads in their effort to recruit new settlers to the West, first for the Burlington and Union Pacific Railroads in the central Plains and later for the Santa Fe and Southern Pacific Railroads.[58] Railroads not only hired consultants and propagandists like Campbell but also funded dry-farming research at experiment stations. The Northern Pacific Railroad had funded the first research on dry farming in Montana, for example, in 1905.[59]

Rather than leave land that was "permanently above the irrigation ditch" out of production,[60] farmers using either the Campbell method or other dry-farming techniques hoped to trap as much moisture in the soil as possible. The first job of a farmer after he claimed his land was to clear it of sagebrush and sod. The plowing that followed clearing was "a fundamental operation of dry farming," turning over the soil to a depth of as much as ten inches. The subsoil plow followed the moldboard plow, breaking through the soil at an even greater depth (ten to eighteen inches) to create a "storage reservoir" for water. Once the soil was sufficiently plowed to allow the rain to enter it easily, the farmer spread a "soil mulch" over the ground to prevent evaporation, an operation accomplished with a harrow. Then he was ready to plant, using a grain drill that left seeds in shallow channels, covered loosely in soil by the chains that dragged behind the drill. The soil mulch was maintained throughout the season by repeated cultivations with a spike or disk harrow, which broke up the surface.[61]

Although dry farming was described as a distinct "method," Campbell and other experts considered it merely good, scientific agriculture applied to drier areas. Campbell relied regularly on the work of USDA soil scientists and agronomists, using the principles they established in addressing the specific problems of the West.[62] The title of his journal emphasized the scientific rather than the "dry" aspects of dry farming, and he acknowledged that his inspiration came in part from the work of Jethro Tull.[63] For John Widtsoe, another dry-farming expert, this method was "in reality farming under drier conditions than those prevailing in the countries in which scientific agriculture originated" in an effort to extend the "great underlying principles of agriculture," which are "the same the world over," to regions previously thought "hopelessly barren."[64] For Widtsoe, too, Jethro Tull's agricultural doctrines were "sound and in harmony with the best knowledge of today" and constituted the "very practices which are now being advocated in all dry-farm sections."[65] Consequently, a full-page portrait of Jethro Tull served as the frontispiece of his manual written for twentieth-century western farmers.

Properly equipped to face chronic drought, or so they thought, farmers scrambled over the dry Plains in the early twentieth century. They recolonized places like western Kansas, eastern Colorado, and the Texas panhandle, where plow agriculture had already been driven out by drought, grasshoppers, or livestock.[66] The land colonized by these farmers tended to be cheap and dry and to produce small yields per acre, relative to cereal yields farther east. Consequently, western wheat farms had to be bigger than the institutionalized 160-acre farm to be worth working. Amendments to the Homestead Act in 1909 and 1910 acknowledged this by enlarging the size of a nonirrigated homestead allotment to 320 acres. The actual size of dry farms was often greater, though, reaching an average of 780 acres by 1925.[67]

But even a parcel of 160 acres required intensive, repeated cultivation with the most efficient contemporary machinery in order to keep the soil in good tilth. The success of dry farming was "closely bound up with the improvements that may be made in farm machinery."[68] Again, just as western dry-farming techniques represented applied agricultural science, dry-farming tools used in the West represented the literal cutting edge of early twentieth-century tillage equipment. Not everybody, of course, was unequivocally optimistic about the success of the great western plow-up. In 1925 Secretary of Agriculture William Jardine wrote that self-sustaining homesteads were possible and practical on the Plains but that farmers had to understand that drought and pests were not the only hazards of Plains agriculture; speculation and debt in commodity production were just as dangerous to farm stability. He stated, "Permanent agriculture must plant its roots around the nucleus of the farm home," providing Plains families' subsistence directly from one-acre gardens before expanding to produce "staple crops for the market on a large scale."[69] But whether individual farm families proceeded with caution or not, dry farming sprawled across the Plains in the 1910s and 1920s with mixed results for farmers.

Population growth and the demand for increased agricultural production during World War I led to higher prices for all farm products, including western wheat, especially between 1916 and 1919. Farmers were encouraged to plow as much as they could and plant patriotically. Then farm prices and overseas demand fell dramatically in 1920 and 1921, creating an agricultural depression well ahead of more widespread economic difficulties.[70] Throughout the expansion of dry farming—systematic and otherwise—adequate rainfall brought success to wheat farmers. The onset of drought in the 1930s, however, sorely tested the reliability of dry-farming methods and revealed the shortcomings of large-scale agricultural development of the Plains.

The explanations for the dust storms that boiled in the air over farms and cities in the 1930s are notoriously contested. In *The Plow That Broke the Plains*, a film made for the Resettlement Administration in 1935, Pare Lorentz unequivocally blamed the plow itself. Oklahoma ecologist Paul Sears published his influential book *Deserts on the March* in the same year, condemning indiscriminate plowing and advocating uses of marginal land that did not destroy its natural grass cover. Grassland historian James Malin argued that periodic drought and dust storms are characteristic of grasslands and the soil types specific to the Plains and that ecologists—particularly those from the East—were entirely out of their element in suggesting anything different.[71] Donald Worster more recently blamed farmers' disinclination to adapt to the drought and traced the arguments of ecologists regarding the history of the grasslands and the much-questioned theory of a natural "climax" of soil, plants, and other organisms that had been disrupted by the plow.[72] Douglas Hurt explicitly recoiled from blaming mechanized agriculture, suggesting that farmers might have used the advanced technology at their disposal more responsibly.[73]

The problem of the dust storms is so complex precisely because it was so multifariously overdetermined. It involved not only plows, climate, and ecology but also the availability and use of equipment and techniques, the failure of absentee landowners to conserve soil and their tendency to abandon land to the wind once a crop failed, the inability of tenants to afford to buy "efficient" equipment and use soil conservation techniques, the scarcity of available credit, and the low price of wheat. Several new federal bureaus were created to deal with particular aspects of the problem, including the Soil Conservation Service (established in 1936 to replace the Soil Erosion Service of 1933) and the Farm Credit Administration (1933).

If the drought had merely curtailed food production for a small rural population, it would have presented a very different ecological and logistical problem. The agricultural settlement of the Plains was predicated on bringing the land under maximum commercial production. Although people had different ideas about what should be produced—wheat, livestock, or a combination of crops and animals—challenging the identification of agriculture with commodity production was not part of the farm recovery or grassland agroecological picture. Assiduous soil conservation and the use of newer, bigger, more efficient machines, bought on credit by those who could get it, might keep more soil in the fields and produce enough wheat to make a living, but the plow and its cousins in tillage could not raise the price of wheat; they could only drive it down faster.

World War II at last brought definitive federal relief because it opened huge new markets for agricultural equipment and produce. Bigger machines available in the 1930s but out of reach to farmers without adequate credit were snapped up in the 1940s; farm prices began to rise steadily with the onset of war in Europe. New tractor gear ratios made tractors travel faster, allowing them to replace work horses almost completely by 1950, and new implements that took advantage of the power of tractors exploded onto the agricultural scene as well.[74] The specter of drought always remained, challenging the Soil Conservation Service, farmers, and equipment developers to contrive ecologically sound methods to negotiate with the soil. But the specter of commodity exchange remained as well. No agricultural bust followed World War II, even though drought returned in the 1950s, affecting a larger area than the original Dust Bowl.[75] The lifting of domestic food rationing and price controls and, more importantly, an expanding food assistance program in Europe and parts of Asia to alleviate shortages caused by the war maintained both prices and demand.[76]

The great plow-up of western wheat land represented one way to force agriculture on an arid landscape. Irrigation represented another, more literally bringing rain to desert lands (and not-so-desert lands) of the West. Like Plains wheat cultivation, western irrigation demanded the reconsideration of the 160-acre homestead. The Desert Land Act of 1877 provided a square mile of land at $1.25 an acre to anyone willing to irrigate it within three years. The prospect of irrigation launched federal surveys of western lands and water sources, including several headed by John Wesley Powell. Once under way, western irrigation demanded the development of new methods of measuring water volume, the codification of new water laws, and the creation of a new irrigating infrastructure, ranging from relatively small ditch diversions of water from creeks and streams, to gasoline engine–powered pumps bringing water out of the ground on individual farms, to large "reclamation" projects involving the damming of major rivers after the Federal Reclamation Act of 1902. Irrigation provided another reason to alienate Native Americans from their land, either by diverting their water to colonists' farms or by flooding their land under reservoirs. Irrigation occasionally challenged farmers to form social arrangements for settling irrigated country different from the every-farmer-for-himself "individualism" of much western colonization.[77]

The project of irrigating the West was contentious and unevenly accomplished, but throughout the process, from the 1870s up to the present, the issue has not been whether the water should be used but how its use should be controlled. Donald Worster eloquently describes the "hydraulic society" that emerged in the West, distinct from the ancient, centralized bureaucracies controlling irrigation and the corvée labor that implemented it in Egypt and China. In the West, agriculturalists' desire to colonize land too dry for eastern-style crop production, combined with their wealth, resulted in a new kind of hydraulic society. Historically it was the state that assumed power over water use, water law, irrigation infrastructure, and the labor that built the waterworks and raised the irrigated crops. But western agricultural interests formed a new class in the hydraulic society, compelling the state on the one hand to subsidize the development of water resources and on the other hand to employ "an anonymous human army," recruited from Mexico, China, Japan, the Philippines, and India as well as from the United States, as "the wage-based answer to the corvée."[78]

The hydraulic society in every manifestation represents a complex and institutionalized enforcement of a literally crushing social hierarchy. But as we have seen, agriculture did not need aridity as an excuse to enslave or proletarianize agricultural workers en masse. In the history of colonization, the hydraulic society of the West simply represented more of the same in terms of property ownership, commodity production, Native American dispossession, and the proletarianization of landless people. For the arid West, the only addition to the recipe for colonization was water.

What was new about irrigation was its use in cultivating enormous quantities of food plants. Noncereal food plants traditionally had been marginalized either in gardens or on small truck farms supplying specific urban markets. Irrigation transformed fruits and vegetables into agricultural commodities in their own right on an unprecedented scale. Garden farming or truck farming had begun this process throughout the country by the end of the nineteenth century, as food-plant production moved out of kitchen gardens—and also decisively out of the hands of individual women—into relatively large commercial gardens or truck farms.

Compared to backyard gardens, truck farms produced vegetables and fruit on a larger scale, and they developed as a direct response to urban growth:

> From the earliest time the quick-growing garden crops have attracted much attention. At first, cultivation of these crops was confined to a restricted area near the habitation because it was convenient to have a fresh supply of appetizing plants and because a large return could be had from a small area. So long as the industries of the people allowed them to occupy the land, and the population was rural rather than urban, the garden formed the great source of the supply of vegetables; but as economic conditions changed and the population of the country became centered in great cities, the garden expanded into an intensive enterprise known as the market garden.[79]

But rainfall and the accessibility of irrigating water largely determined where commercial vegetable-crop production could expand in scale or productivity. Market gardens were generally out of the question on the Plains, which had already been pressed into service as the American Breadbasket. In the East and the South, fruits and vegetables could be grown without irrigation, but a horticultural expert wrote in 1940 that "within recent years irrigation has been adopted to supplement the rainfall during the growing season and has been an important factor in the progress of vegetable-crop production" in those regions.[80] In the West, irrigation was a supplemental strategy in raising grain and forage crops but was essential in raising fruits and vegetables on a large scale. Commercial horticulture in the West began only with the use of irrigation and was entirely dependent on it. And yet the success of the western fruit and vegetable industry was remarkable.

California emerged as the greatest fruit- and vegetable-producing area in the country. The national aggregate of fruit production stopped expanding in 1915, and vegetable production stopped expanding in the 1920s. But in California, the acreage planted in fruit trees increased by half between 1909 and 1925, and the acreage planted in vegetables increased more than ten times between 1909 and 1936. By 1940, California produced almost half of all of the fruit consumed in the United States, 90 percent of the dried fruit exported from the United States, and 70 percent of the processed fruit as well as a great deal of lettuce, tomatoes, asparagus, and spinach. The value of California's fruit and vegetable crops was about $300 million in 1936, almost three times more than the produce of any other state.[81] Other major western horticultural crops included winter lettuce from Arizona, summer lettuce from Colorado, cantaloupes and muskmelons from Arizona and Colorado, onions and cabbages from Texas, apples from Washington, and strawberries from Oregon.[82]

Like bonanza wheat farming, the fruit and vegetable bonanza in California had certain social and economic consequences. Unlike the wheat boom or the livestock boom, however, even by 1940 no historian had "recorded even the outlines" of the Texas Onion Rush, the Northwestern Apple Rush, or the Asparagus Rush. But the contours of these rushes were familiar: "Many who rode the crest of the waves made fortunes; others who were caught by the undertow of receding prices met disaster."[83] And, like bonanza wheat farming, California fruit and vegetable growing took place on ever-larger farms in the hands of ever-fewer landlords, many of them absentee. In 1916, 310 California landowners controlled more than 4 million acres, which might have supported as many as 500,000 additional residents on small farms.[84]

California was socially stratified by the wealth of landowners, the size of their holdings, the fact that the volume of cheap produce set the market conditions for other farmers, and the dependence of large farms on large numbers of laborers. Mexican workers in particular had been recruited to work in western fields in the 1890s as agriculture expanded dramatically in labor-intensive and seasonal production of fruit and vegetables as well as hops and sugar beets. Women and children worked in the fields as well, though unpaid. Mexican workers became indispensable in western fields before the turn of the century, doing work that Anglo workers would not tolerate—until the Depression, when Mexicans and Mexican Americans were fired to make room for Anglo workers.[85]

Their indispensability eventually led to the bracero program of the 1940s. Growers' organizations during this time contracted Mexican men by the tens of thousands, establishing a pool of low-wage workers to be mobilized at peak harvest times or simply held "in reserve" and "stockpiled" in the absence of immediate need.[86] The bracero program institutionalized a system of "semicaptive labor," as historian Erasmo Gamboa put it, which included all of the abuses associated with such systems.[87] Interestingly, it was cotton production in California (as well as in Texas and Arizona) that provided a relatively early excuse for Mexican exploitation, before irrigation allowed the southwestern landowners to diversify into fruits and vegetables.[88]

Donald Pisani documents early protests against land monopolies in California but notes that the belief that farmland should be held in relatively small parcels by individual families was partly based on a desire to remove industrially essential but socially undesirable laborers, particularly Asians, from the fields or to allow soldiers to establish farms after World War I. Moreover, reformers protested the labor situation, not because of harsh working conditions or the impoverishment and exploitation of landless people, but because "ignorant and nomadic farm labor is bad." According to one propagandist, the goal of California land reform was to "make the farm as attractive as the office or factory for men and women of character and intelligence," perhaps not so much for their own happiness but in order to release California agriculture from dependence on a large non-Anglo working class. Nevertheless, California agriculture became permanently and increasingly dependent on Mexican and Mexican American workers.[89]

One of the technologies that aided the expansion of the horticultural-crop market was canning, which allowed crops to enter the world market.[90] Fruit and vegetable canning had been an American specialty since before 1860, but the cans were handmade and relatively expensive. In 1868 handmade cans gave way to cans cut and shaped by machine. Big canneries emerged in Chicago, where so much of American agriculture found its greatest industrial and commercial expression. The industrial packing of fruits and vegetables made sense only in the context of an agricultural sector already moving toward mechanization and long-distance transportation, encouraging developments in both harvesting equipment and food refrigeration and preservation.[91] Fresh as well as frozen foods traveled east in increasing volume through the 1940s. Per capita consumption of processed vegetables more than doubled between 1915 and 1950, and most vegetable crops were available to "consumers" year-round.[92] Again, California stood out in the production of canned goods; between 1939 and 1950, California canned more fruits and vegetables than any other state. Mexican women provided much of the labor necessary to keep these industries running, especially after 1900.[93]

Commercial fruit and vegetable production, supported by new branches of agricultural science, was clearly understood as the modern replacement of the kitchen garden, which appeared to have become unnecessary and impractical for a largely urban population. Two truck-crop specialists wrote in 1928: "With increasing specialization, the development of practices suitable for large-scale operations, and the increasing volume of production, . . . most of the ancient aspects of 'gardening,' as we commonly understand them, have disappeared from commercial vegetable growing."[94] Two others wrote in 1940 that horticulture was "one of the oldest arts" but was only "a young science." Historically, "horticulture" referred to what happened in the garden rather than the field. "Obviously," they wrote, "this distinction, although satisfactory during feudal times, is no longer valid, for many single horticultural crops are now produced in larger areas." Horticulture "is much more than gardening," they continued. "It is an industry and occupies important economic and cultural positions in the field of agriculture."[95] Horticulture had literally escaped from the garden into the "field of agriculture" by 1940. Food crops, which for many centuries had been outside the main agricultural business of grain culture, were absorbed into domestic and foreign markets as commodities, into agricultural science as new objects of inquiry and improvement, and into the overall project of western colonization.

The transformation of the "garden" into the market garden and the monocultural vegetable field raises the issue of "gardening" as distinct from (and largely replaced by) field agriculture. The identification of gardening with women, as opposed to the field agriculture identified with men, has attracted several scholars hoping to describe gardening as a site of alternative agri/cultural values.

Annette Kolodny identifies the garden as a major trope in white women's imaginations of the West, very different from the "virgin land" awaiting the plow that men imagined. She wrote in her preface to *The Land Before Her*: "Massive exploitation and alteration of the continent do not seem to have been part of women's fantasies. They dreamed, more modestly, of locating a home and a familial human community within a cultivated garden."[96] Kolodny was writing primarily about women's *ideas* of the West as a garden, not gardening in the West as a material practice; her focus was therefore on women's impressions of the landscape and their expressions of the desire to garden. She wrote: "Given the choice, I would have had women's fantasies take the nation west rather than the psychosexual dramas of men intent on possessing a virgin continent. In women's fantasies, at least, the garden implied home and community, not privatized erotic mastery."[97] Although "fantasies" (including ideologies) have a great deal to do with material manifestations of any cultural project, what seems less clear with respect to women's gardens is the extent to which their fantasies, by the seventeenth, eighteenth, or nineteenth century, found expression in the material practices of their gardening or what relationship the fantasy of the "garden" might in fact have with the fantasy of the "virgin land." The women's fantasies that Kolodny studies may represent a form of cultural memory carried from a much more remote past without an accompanying material, practical memory.

Vera Norwood provides a history of women's gardens in "Designing Nature: Gardeners and Their Gardens" in *Made From This Earth*. Many of the gardens she presents are ornamental, museumlike specimens owned by the well-to-do, who have the "grounds" to spare on such projects—the most likely gardens to leave records of both their design and the philosophy of their gardeners. "Vernacular" gardens, she notes, have been studied primarily by people interested in garden design rather than plant choice, a lapse that will hopefully be corrected before long. She describes women as central in movements to wrest control over patches of land for urban gardens. For Norwood, women's gardens of all classes embody the fact that "a love of gardening pervades women's culture,"[98] although she leaves the question of how women's culture is associated with the home unasked. Gardening involves "conserving green spaces" and is "meshed with domestic roles," which include beautification, enclosure of space for privacy, and the culture of domestically useful plants in kitchen and herb gardens.[99]

This picture of domesticity is predicated on the separation of masculine and feminine spheres that robbed women of most of their productive expertise and authority. The development of women's expertise as ornamental gardeners among the wealthy perhaps reflects more the exaggerated leisure of their class, which allows them to engage in an assiduously nonproductive production of beauty as part of an obligation to display their wealth and "cultivation," than any innate love of horticulture. Among the less privileged, the control of urban space and the production of both beauty and utility in small gardens suggest the degree to which domestic production has been marginalized, relegated to what land can be removed from the authority of city governments, pressed into use in the yard, or spared near the house on a farm. The smallness of these gardens, relative to either cities or fields, has to be seen in the context of women's exclusion from control of these larger areas rather than an expression of some essentially feminine appreciation of small, enclosed spaces.

Both Norwood and Kolodny take the scale and uniqueness of women's gardening for granted as a beginning point for feminist analyses without focusing substantially on women's tools, women's crops, or a comparison of Euro- and African American women's gardens and other women's horticultural/agricultural production. Much more work needs to be done relating the ideological history of women's gardening to their material practices of gardening to determine what forms of "gardening" can be offered as alternatives to field agriculture.

Gardening in the West was never seen as more than a supplement to the income gained from the fields, even if it provided the means of subsistence for a family improving a homestead. In this way, it was analogous to twentieth-century farm women's "off-farm income," something extra to help the family, and the farm, get by.[100] The garden was frequently near the house since it was "under the care of the farmer's wife," but at least one expert recommended planting some of the "heavy" food crops, including potatoes, melons, squash, and cabbage, in the fields, where they would be cultivated exclusively by men. All of these crops, except for melons, represented staple foods that were to be stored and consumed by the family all year. Gardens near the house, in that case, would grow small root crops like carrots and onions and summer vegetables, further restricting the responsibilities of "the farmer's wife" beyond subsistence food production to raising only food eaten fresh from the garden.[101] Other gardening experts failed to acknowledge that women had even that much responsibility, referring repeatedly to the gardener only as "he."[102]

In the early twentieth century, the value of the garden was greatest on farms in the South, where as much as 82 percent of a family's food was produced in the garden. In the West, gardens produced between 50 and 70 percent of a farm family's food, and in the East, rural families raised about half of their own food.[103] Although always valuable, kitchen gardens became less crucial to a family's survival as other food sources became available—canned food, first, and later refrigerated and frozen food. One gardening expert wrote in 1916, "Many farmers nowadays seem to feel that it is beneath their dignity to bother with a garden." The contemporary farmer instead "keeps his garden on a grocery shelf."[104] As large-scale western fruit and vegetable production expanded, and the industrial canning process and long-distance transportation were perfected, the image of a farming man's garden "on a grocery shelf" was quite apt.

Western gardens looked a great deal like the fields and were cultivated by men and boys with the equipment used in the fields.[105] They were planted in long wide rows, following the recommendation of Jethro Tull, so that "most of the cultivation could be done with a horse." Some experts believed in maximizing efficiency in the garden as well as in the fields, suggesting that the only difference between the two was that a garden-field was closer to the house and raised specialized crops. Since there was plenty of land on a farm, wrote one experiment station gardener, "economy of time is of more importance in the case of the garden than economy of land." The garden he recommended was 280 feet by 77 feet, much of it empty space.[106] According to one prescription, the garden should have a wagon-sized gate for bringing in manure and "the larger implements of tillage," and it should be as convenient to the barn as it is to the house.[107] Another specialist acknowledged that not all gardens would be cultivated by horse-drawn equipment but stated that those that were should be laid out in long, straight rows. Gardens tended by hand might look "quite different," arranged in small sections surrounded by walking paths. He recommended horse cultivation, however, to save labor.[108] This advice remained generally current into the 1930s, when the same expert published a bulletin practically identical to the one quoted above, which was published decades earlier, advocating the same garden design, crops, and tools.[109]

It is possible that hand-cultivated gardens were, by the late nineteenth century, associated with immigrants, particularly those from countries judged to be poor or less socially or technologically "advanced" than the United States. One USDA garden expert wrote in 1906 that Mediterranean immigrants in California were cultivating successful gardens by hand. They made small depressed or raised beds in terraces on steep hills, often irrigated, which required an "immense amount of hand work," but he added that "the Mediterranean immigrant seems born to it." He also described Spanish methods of irrigated gardening, in which gardeners planted the ditch banks that led to the fields.[110]

Although this information about garden culture is taken from prescriptive literature and thus probably obscures the existence of a variety of garden forms, the fact remains that western fruit and vegetable industries were in the process of substantially changing Americans' relationship to food crops. The emphasis on greater efficiency of labor in the garden merely recapitulated what was happening on a larger scale in the produce fields of the West. One of the last components of domestic food production actually controlled by the individual gardener—seed saving—was replaced during the second half of the nineteenth century by commercial seed production and marketing. Herbs, those commonplace denizens of women's gardens since time immemorial, were increasingly replaced by patent medicines. And as implement manufacturers turned their attention to garden tools in the 1880s, they directed their advertisements at men and boys, not at women.[111]

The garden and the expertise it required may represent a principle of usufruct absent from European agriculture for centuries, but the western garden had more in common with the fields than seekers of "alternative" agricultures might wish. One of its striking features was the fact that the woman who tended it was not a farmer herself but the wife of a plowman. It was this domestication of women within their own society that also prevented federal agents from seeing Native American women as farmers.

These changes in American agriculture—which underscored again and again the emphasis on "efficiency," commercial value, and greatness of scale—as well as the accompanying marginalization of noncommercial forms of agriculture and the arrangement of women, Native Americans, and farm laborers in their respective places in an elaborate social hierarchy represent choices made by the powerful and bargains made by the less powerful. Nowhere has it been predetermined that "man" must maximize "his" agricultural efficiency and production. The choice to abandon the hoe for the ard and the ard for the moldboard (and so on, up to the diesel tractor pulling an eighteen-foot drill) is not a natural one. But that is precisely how agricultural history tends to be understood. The agriculture of the West spread out on an unprecedented scale and epitomized the commercial and social values inherent in field agriculture for many centuries. It is no accident that such "development" took place against the backdrop of wilderness, the "virgin land," Nature. According to anthropologist Renato Rosaldo, "development" takes place when we permanently transform something understood as "primitive" or natural, mourning at the same time the past that has been destroyed in the name of progress. The whole process is understood to happen naturally. Pierre Clastres similarly observed that "*biological* metaphors are invoked" in accounting for the development of the state from an allegedly primitive origin.[112] Rosaldo further refines this analysis: social development is analogous to the maturation of a child, who, after becoming an adult, looks back on the inevitably lost childhood with nostalgia.

Rosaldo ascribes this understanding of history—both of the child and of society—to imperial subjects, who have naturalized their subjugation of other people and places for their own colonial gain. Imperialist nostalgia, as he calls it, allows colonizers to justify their mistreatment of other people and colonial landscapes by naturalizing the process of development as a regrettable (because it destroys the "native" and nature) but inevitable process of natural growth.[113] Agricultural history in general, and the history of agriculture in the West in particular, is saturated with this nostalgia. There is a good reason for this. Agriculture represents the action of culture against nature, which defines both culture and nature simultaneously. Agriculture is the permanent transformation of a set of subordinate beings forever after identified with "nature"—plants, animals, laborers, soil, water—into something domesticated and the relentless search for means to expedite this transformation.

Nature in a sense was created by the plow and culture. Nature was, of course, female and came into the English language at the same time as the plow, in the eighth century, replacing the woman who produced food *in* the field with the woman *as* the field. Women were designated as "wives" at about the same time; they were not understood as autonomous but subjected to a patriarchy that would soon name the "husband" as head of household, also aptly describing an entrenched plow culture that gave men the power of cultivation. Woman was the literal mater/ial from which the cereal crops were to be raised; the plowman was her husband. She no longer controlled production herself. "Culture" appeared during the great reclamation of the sixteenth century and meant "agriculture," although in a few years "agriculture" was a word of its own.[114]

Conceptually, one is born(e) by or from the body of nature, not within it; the idea of nature places us already at one remove from it, beyond it, born. This state of affairs implies an inescapable (indeed natural) history. The word "culture" names what happens in this history of being born and then separating from nature, improving upon it, and bringing its latent potentialities into being through certain kinds of work in the arts and sciences. Culture exists where nature has been permanently altered, even obliterated, where a society measures its worth by its distance from "nature"—from its birth—all traces of which have been relegated to the past. Nature makes culture possible, and a "culture" looks back with a certain nostalgia to the nature that gave rise to it, at the same time that it is only through the transformation of nature that a culture understands itself as "advanced."

Cultivation, or the arts and sciences of improving nature, is an act of transformation that takes "wild" territory—virgin land—and breaks it as one would break an animal or subjugate a slave, processes, incidentally, accompanying many agricultures supported by states and empires. It is a process of domestication by which a plowman enforces his domination over cropland in such a way as to render the land permanently "improved." The permanence of his efforts may have been compromised by hard times, but more often, and certainly in the American West, it was supported by markets, railroads, and state-guaranteed property. Land not under the plow was not improved and therefore appeared to the plowmen to be unoccupied.

As we might expect, the process of domesticating the soil is often described as an act of violence, which indicates both the will to mastery implied in cultivation (as such) and the intractability of "nature." The distinction between primitive and advanced agriculture often hinges on the extent of "breaking" that tillage accomplishes. Alfred Crosby describes the plow as "a disruptive, even violent instrument."[115] The moldboard plow was a progressive improvement on "the earliest plow," which was "essentially an enlarged digging stick dragged by a pair of oxen."[116] The most common property of these lesser implements seems to be the fact that they "merely stir" the soil: "The first agricultural implement used by prehistoric man . . . was a hooked stick, or sometimes a stag's horn, adapted to the work of digging and stirring the soil in planting seed. This rude tool—it could scarcely be called an invention—developed in course of time into something more like a plow, the forked stick with a branch to which animals were attached."[117]

The distinction between civilized and primitive agriculture was so deeply entrenched that the most outspoken critic of the moldboard plow, Edward Faulkner, used it to his rhetorical advantage: "One of the persistent puzzles [in scientific agriculture] has been the fact that an ignorant, poverty-stricken [twentieth-century] Egyptian who stirs the soil with his ancient crooked stick can produce more per acre than his British neighbor whose equipment is right up to the minute."[118] But, of course, colonial rulers, including Americans, forcibly encouraged the natives, wherever they found them, to adopt the plow. In 1899 American agricultural boosters believed that the "Filipino will gladly exchange the crooked timber he stirs his rice bed with today for an American plow. . . . There can be no doubt that with the development sure to follow under American rule, a demand will come for American implements."[119] Ancient plows were built "only for the purpose of breaking [into] and stirring the soil, the bottom having been invariably a simple wedge, with no power to turn a furrow."[120]

This issue of power is crucial to the technology of plowing because without it an agriculture slipped down the scale toward the primitive. A historian of Australian agriculture explicitly casts the problem of power in terms of "man's challenge to the soil" in his "battle to conquer" it. For this historian, the aboriginal's digging stick (wielded in an illustration by a girl) is the most primitive and universal agricultural tool, beyond which agrarian people progressed "to a hoe, fork, spade, plough, and harrows for their cultivating tools" and thereby presumably won the battle.[121] For Lynn White, the moldboard plow is a "far more formidable weapon against the soil than is the scratch plow," its action an attack that "handled the clods with such violence that there was no need for cross-plowing."[122] That inferior agricultures were associated with a state of nature is clear from the remarks of an American visitor to Russia in the 1890s. He noted that Russian furrows were only cut and were not turned completely over and that a boy followed the plow, using a mallet to break up the clods of soil. After that, the ground was harrowed repeatedly. The visitor concluded his observation with the hope that "our country will never be reduced to a condition necessitating such methods," but he added that Americans could nevertheless learn a great deal by observing "the methods of these children of nature."[123]

The spectrum of agricultural implements, from digging sticks to successively powerful plows, is often presented as a historical "progression" when in fact it represents the range of agricultural possibilities operating at any historical moment. The progressive understanding of tillage precisely complements the history of agriculture outlined by Isaac Newton, which began with hunters and gatherers and ended with western wheat.

Likewise, agricultural changes that took place in the West were understood as improvements on the agriculture of the East. Robert Ardrey and Walter Prescott Webb, for example, made a distinction between eastern and western farming, between old and new farming, on the basis of an alleged shift away from subsistence agriculture. Wooden moldboards went the way of ramshackle small farms in an era of agricultural expansion and mechanization. But wooden moldboards were not inherently backward, and a forty-acre farm was not merely producing a subsistence for a family. As we have seen in the case of the Pima, even an ard could prove to be a satisfactory implement well into the nineteenth century (and no doubt into the twentieth century among other people). The issue for Webb and Ardrey seems to be one of scale as well as subsistence. Webb pointed out that the demand for the biggest equipment available came from the West.[124] Ardrey wrote that older, eastern farming "was conducted on a far smaller scale then, for the cities being small and few in number, the market for farm products was limited, and the average farmer contented himself with growing enough for his family, with a small surplus for purchasing the very few articles of commerce indulged in at that early day."[125]

By contrast, the raison d'être of the modern farm was to produce cash crops, and to eulogists of American agricultural progress—those commentators like Ardrey who equated development with increasing efficiency and durability of machinery—western farming was the epitome of modern agriculture. Large fields and efficient machinery did away with the few elements of subsistence farming that remained in the East and brought agriculture into the industrial age. It was the steel plow that made this agriculture possible. Without it, prairie "plowing was done with great difficulty, and in some sections it could not be done at all with the old style plows, except under favorable conditions. The new kind of steel was like oil upon the troubled waters, and proved itself worth millions annually to the farmers of the west."[126]

Agriculture in general represented the "breaking" of nature into culture, and the vocabulary of breaking became more closely associated with the modern steel plow than with any other implement. The purpose of these plows was to sever the roots of native prairie plants and turn the slices upside down to kill the sod. The point of breaking the prairie was not simply to break into the sod but to replace it. Eastern and far-western plows turned soil on land that had been cleared of forest by axes. Western breaking plows alone were the transformative tools in the domestication of western grassland. Just as the forest was "virgin" timber ready for the axe, the prairie was "virgin" as well, its sod "wild" and undomesticated.

An expert on farm machinery in 1911 distinguished between plows with straight moldboards "used for breaking wild grass sod" and those with shorter, curved moldboards used to plow "tame grass sods." He continued: "A plow [that turns wild grass sod] is known as a *breaker*."[127] The steel breaking plow, identified sometimes as a prairie breaker, "was a most important aid in the agricultural occupation of the prairie. In the first place, the presence of a variety of tough roots beneath the innocent-seeming surface of sod made a sharp cutting edge an all important desideratum in plowing."[128] This writer's use of the phrase "innocent-seeming" underscores the variety of cultural forces at work in descriptions of plowing the prairie that are associated with agriculture more generally. Like a "savage" society, wild sod only appears to be innocent, whereas in reality it represents a dangerous force of nature that must be domesticated.

Many new tillage implements had a place in western agriculture, especially dry farming. But whatever tools might have been used after the fields had been cleared for the first time, the breaking plow was always the tool used to bring a new piece of ground into agricultural production. As late as 1948, the breaking plow was still the necessary precursor to any further cultivation. By this time, single-bottom plows had long ago given way to gang plows, and horses and oxen had given way to the tractor, but a farmer still needed a "breaker bottom . . . for turning virgin sod and land being returned to cultivation after years of idleness." Other plow-bottom shapes were suitable for turning "tame grass sods."[129]

Western agriculture was ultimately about cultivating the nature that was the American West in the minds of the colonists, domesticating that bountiful and barbarous wilderness. What happened in the West was a more recent variation on the theme of colonization that had begun with plows long before. Even the "savages" had antecedents in the warlike or recalcitrant Celts and pagans and other roving or "backward" folk. The American West is not even a unique example of the rapidity of change brought about by plow agriculture. Since the "Green Revolution," plow agriculture and commodity production often represent a catastrophe in local food production in the so-called Third World.[130]

What is fascinating about the western plow is the fact that it was identified with civilization even when agriculture ceased to be the primary occupation of Americans, just as farming itself became identified with "nature," a relationship that a nineteenth-century wheat farmer would have found shocking. This is the imperialist nostalgia that lies at the heart of all of the components of western agriculture, from the clearing of the forests (and the preservationists' horror at having cleared them), to the establishment of purebred livestock industries on grasslands perhaps never touched by the plow. Agricultural specialists in every field transformed the "wild" into the "tame." The violence of this process was most vivid regarding those elements excluded from the domesticated scene, including Native Americans and, as we shall see, weeds. European women had been domesticated within this system perhaps millennia before it was turned against Africans, Caribbeans, Australians, and North Americans. The issue of "women's culture"—both women's society and women's agriculture—is riven by hierarchies as a result. As farmers attempt to hold onto their land in the face of urban sprawl, cheap commodities, and expensive equipment, agriculture itself has now become naturalized as the antecedent of "industry," eclipsed by a nostalgia made possible through its own relentless work of culture. Not one atom of this history was inevitable, regardless of how naturalized it has become.

Grass

We see a great drive which employs a *grand number* of small instruments.
—Friedrich Nietzsche, *Philosophy and Truth*

Like forests, grasslands have a place in the history of civilization. Somewhere between the stages of hunting and gathering and sedentary agriculture, grasslands become the scene of pastoralism as human beings wend their difficult way toward civilization, so the story goes. In the American West, where it has been popularly supposed that civilization had to start anew at the edge of every frontier (independent of Frederick Jackson Turner's expression of the idea), pastoralism represented an intermediary stage in the agricultural settlement of territory. In the case of the western grasslands, where the level of precipitation discouraged eastern-style agriculture, this intermediary stage became the primary expression of colonial occupation. One could make a civilized fortune running cattle or sheep on the range, though at the expense, of course, of the people and animals who had lived there before.

What was backward about open-range, nineteenth-century stockraising was not the pastoralists themselves but their lack of scientific method, their failure to ensure the perpetuity and improvement of this form of occupation: they were not sufficiently agricultural. Like western forests, western grasslands were eventually lifted out of their relatively primitive stage in the progress of American civilization to become fully agricultural places. This agriculturalization of the grasslands made it necessary to look back at the period of the open range as a less systematic form of occupation, less scientifically and commercially sound, and in that way a more primitive phase of an otherwise sophisticated expansion of the colonial American state over western territory. Like the forests, the grasslands had to be absorbed into the general process of cultivation central to American colonial occupation. The sciences of animal husbandry and nutrition, agrostology (the botanical study of grasses, analogous to dendrology), and eventually range management accomplished this agricultural transformation of colonial pastoralism.

The pastoralism of the open range, from the 1860s to about 1890, had a number of things in common with migratory lumbering of the same era. Cattle and sheep, like trees, were primarily valuable in number only. Western stockmen did not pay much attention to the quality or breed of the animals themselves until the end of the open-range period. Also, cattle and sheep were driven wherever there was grass, regardless of boundaries between public and private land. They grazed until the grass was gone and then were moved elsewhere. Unlike trees, grass could grow back relatively quickly, but the quality of the range deteriorated nevertheless, as perennial grasses gave way to annuals and native plants gave way to introduced weeds. Not only did the overgrazed range provide less nutrition for wild and domestic animals alike, but overgrazing had compromised the legendary immortality of the grass. Some places were entirely barren. When bad winters and drought-depleted, overgrazed ranges broke up the cattle boom in the 1880s, stockgrowers were compelled to change their methods in order to remain commercially viable, and the USDA was ready to support research to develop more "advanced" forms of western pastoralism.

Semimigratory practices like long stock drives gave way to fenced pastures and feedlots; "scrub" stock was replaced by "improved" stock descended from purebred animals; and the grassland range became the subject of new sciences, beginning with agrostology in the nineteenth century and culminating in range management in the 1920s. The transition from destructive and "primitive" open-range stockgrazing to scientific range animal husbandry and range management was an attempt to guarantee the stability and profitability of western stockraising, just as forestry promised to stabilize lumbering and ensure the permanence of the forests.

Again like forestry, the new sciences of range plant and animal management did not limit colonial control of state territory but resulted in a proliferation of objects to be managed for maximum profit or production. Whereas in open-range stockraising, only the number of animals and the acreage of range available were crucial in claiming territory and making money, a scientific range industry controlled the breed, size, shape, and internal qualities of individual animals as well as the area, type, and reproductive cycle of the vegetation (native or otherwise) to be grazed. Eventually, range management included nondomesticated animals in its research agenda as well.

The cattle boom of the 1870s and 1880s grew out of the extraordinary surplus of Texas cattle, descendants of Spanish cattle, many of them feral, that had multiplied in Texas during the Civil War. Cattle were driven north to herd buyers and markets, creating cow towns on railroad lines and a theater for the culture of the cowboy, not to mention fortunes for investors. Sheep were generally as numerous on western ranges, if less glamorous than the lanky, indestructible longhorn. These sheep were also of Spanish origin. Although trailing sheep never achieved the iconographic status attached to roping steers and driving cattle, a "sheep kingdom" grew up as well in the last few decades of the nineteenth century. Conflicts between sheepmen and cattlemen over range territory were sometimes violent and sometimes indirectly damaging, as when stockmen allowed animals to overgraze an area so that the enemy would find it bare.

Although the bison had been a mainstay of Native American Plains economies, the fact that it was a grazing quadruped occupying a niche more profitably filled by European animals guaranteed its near extermination by 1876. Cattle and sheep grazed these deterritorialized ranges from Mexico to Canada. In 1854 one observer praised the lush grasslands of Texas; by 1883, these lands were overrun with livestock, and the grass had given way to mesquite brush and prickly pear. As early as 1865 in California, another observer noted the disappearance of the native grasses.[1] Perennial grasses succumbed to trampling and overgrazing, and weedy annuals became established instead, which provided less nutrition for wild and domestic animals alike.

Since cattle demand greener pastures than sheep—sheep will eat weeds and brush that cattle will not—the depletion of the range was felt first by cattlemen. During the hard winter of 1886–87 in the North, cattle faced blizzards and severe cold on drought-stricken and depleted ranges and died in great numbers. Meanwhile, "nesters" were already moving west armed with new techniques for dry farming and irrigation. They claimed the open range for crop agriculture and eventually closed it, with the help of barbed wire, to trail shepherds and cattle drovers.[2]

Federal homestead land-granting policy after 1862 had encouraged western irrigation and crop raising. The homestead acreage was increased again by the 1916 Stockraising Homestead Act to a full square mile, or 640 acres. These changes allowed ranches to enclose property under private ownership, securing water and forage-crop fields that would be supplemented by seasonal grazing on the public domain, often in national forests. "Family-sized" operations on land where it might take eight acres or more[3] to support a single cow and her calf would eventually occupy from four to twenty square miles.[4]

The primitive nature of open-range stockraising is generally expressed in two ways: open-range stockraising is described as part of the process of colonial settlement, which is supposed to culminate in sedentary agriculture, and as lacking systematic or scientific methods. If we stand at the Cumberland Gap with Frederick Jackson Turner to watch the pageant of American civilization, the buffalo disappears first, followed by the Native American, fur trader, hunter, and cattleman in that order, before the pioneer farmer comes "and the frontier has passed by."[5] Ray Allen Billington, a Turnerian partisan, repeated in 1959: "On the cutting edge of the moving frontier roamed fur traders and trappers. . . . Next, when conditions were ripe, came the miners. . . . Not far behind were the cattlemen, constantly seeking grasslands where their herds could roam freely without the hindrance of farmers' fences." Other pioneers followed, either to buy or settle on land or to sell goods and services to permanent settlers. The presence of these residents and investors "signaled the end of the pioneering period."[6] Agricultural historian Gilbert Fite describes the progress of settlement in similar terms, with a "leather-clad mountain man or a gun-slinging cowhand" preceding the "overalled farmer" on the frontier. The trappers and drovers hoped to "hold civilization back, to protect the beaver streams from intruders, to guard the grasslands from the plow." Fite characterizes the early frontier uses of the land as friendly to "raw nature," in contrast to agriculture, which was a form of domination over nature and the sign of "advancing civilization."[7]

Given the depletion of grasslands as a result of open-range stockraising and the elimination of the bison, which made stock operations possible, it is difficult to see how the "cowhand" was friendly to raw nature. This was true only in the sense that open-range stockraising did not seek to improve nature by replacing the grass with crops. It claimed and held territory (while making a profit) in the subhumid West that in the 1860s and 1870s could not be claimed directly by sedentary agriculture. This colonial imperative to territorialize the grasslands resulted in a violent and rapid deterritorialization of a preexisting Plains economy and ecology, long before Euro-Americans knew how to farm the western grasslands with even marginal success. From this point of view, open-range stockraising was primitive because it merely reserved an area for a postponed cultivation.

But open-range stockraising included profoundly agricultural elements, more than did migratory lumbering. Something with no commercial value was improved and transformed into something with a great deal of commercial value, as grass was converted into beef, mutton, wool, and hides. Raising stock was a way of profitably exploiting otherwise useless territory on a large scale.

Also, open-range stockraising used the bodies of domesticated animals rather than "wild" ones as much as it depended on the natural range. In these ways, open-range stockraising was already a form of agriculture, but it was "primitive" in the sense that not all of its elements were systematically subjected to the forces of "improvement." Progress in this type of agriculture consequently took the form of emphasizing those elements that were already agricultural (the productivity of grass and domesticated animals) and eliminating that part of western stockraising held in common with noncolonial pastoralism—its nonsedentary use of the range.

Texas was the first site of the cattle boom, since it was where the cattle were. Texas entrepreneurs hired cowboys to gather and drive cattle over trails on the public domain to pasture and to buyers. Abilene became the first major market for cattle, lying at the point of intersection between the south-north route of the cattle drives and the western terminus of the railroad. As the railroads expanded westward through the 1870s, the center of the cattle market shifted west also. The grazing range, meanwhile, continued to move north as increasing numbers of cattle filled otherwise "understocked" grasslands. Northern entrepreneurs bought Texas cattle to stock the ranges of Colorado, Wyoming, and Montana, but they bought purebred eastern animals as well and were among the first ranchers to experiment with herd improvement. Since there was no ready supply of "wild" cattle free for the taking on northern ranges, cattle already represented an investment to stockmen who bought Texas herds to fatten and sell for beef. Fatter cattle, of course, brought better prices in Chicago, so northern stockmen had an added incentive to raise their own herds of improved cattle rather than simply buying and feeding thin Texas stock.

Once purebred eastern cattle had been introduced on western ranges from Texas to Montana, they were all susceptible to "Texas fever," a tick-borne disease carried by longhorns but to which, not surprisingly, these hardy cattle were immune. By 1880, the northern ranges were stocked as heavily as those in Texas as more and more investors from as far away as Great Britain sought their fortunes. Together, the improvement of herds, the threat of Texas fever, and the competition of herd owners for range territory all encouraged the enclosure of the range and the end of the great cattle drives north from Texas. Quarantines were imposed in an attempt to stop the movement of Texas cattle as early as 1859 in Kansas. These efforts continued as cattle were moved north by rail as well as on foot through the 1880s. Barbed wire made its commercial debut in Illinois in 1874. Demand for this practical material grew tremendously through the end of the decade. Faced with the loss of a major market in the North and dependence on the railroads rather than trails to transport their stock to market, some Texas stockmen urged the establishment of a national stock trail in 1884. This unfenced corridor would extend along the public domain from Texas to Canada and ensure a market for tick-carrying but otherwise healthy south Texas longhorns. The measure was defeated in Congress. The national trail lost out to other Plains states' quarantine laws, and the general trend toward enclosure, herd improvement, and shorter drives to local railheads continued.

In any case, the boom had reached its peak. Overstocked ranges throughout the Plains could not support large concentrations of cattle once the grass was depleted by grazing and drought, and during the winters of the mid-1880s, thousands of animals perished. Investors liquidated their remaining stock, often to pay creditors, and prices for cattle plummeted by the summer of 1887. The cattle raising that persisted had to take place without large outside investment or the concentration of animals built up during the boom and without dependence on the open range. As Walter Prescott Webb put it, "The haphazard, free-and-easy methods of ranching must go with the native cattle and the free range that produced them."[8]

Of course, the cattle were not native at all, but Webb's use of that word underscores the agricultural aspect of range stockraising. The otherwise uncultivated grass became a crop in the bodies of animals subjected to all of the energies of "improvement" associated more generally with agriculture. Nineteenth-century pastoralists diligently cultivated "wild" Texas cattle and scrub sheep by breeding them "up" to purebred bulls and rams. The longhorn's "wild nature," according to one historian, "was considerably subdued by crossbreeding with Herefords, shorthorns, and other more sedate eastern cattle." Longhorns exhibited all of the bad behavior of other "wild" beings, luring stock away from "tame" herds, "spiriting them into the mesquite jungle, and generally making a nuisance of themselves."[9] These cattle were "almost as wild as the buffalo"[10] and just as capable of finding water and feed on dry grasslands. Mari Sandoz wrote of the longhorn cow: "She could move fast enough if the rains failed and the creeks and water holes dried up. Sticking her wet nose into the wind, she followed it to other holes or other creeks, and when they were gone, too, she pursued the smell of moisture under the dry stream beds up to the canyon springs or down to surfacing water. If necessary she could cross dry tablelands to other watersheds, even those she had never seen before, even if they required two, three days of hard, dry travel." She was, "next to the buffalo out on the open plains, the first to know when snows and blizzards, the blue northers, were due."[11]

Longhorns may have "grown wild" during the Civil War, when Texas stockmen were otherwise preoccupied,[12] but there is an implicit consensus among Anglo historians of animal breeds that Spanish livestock was more "wild" and also less valuable individually than northern European stock. The longhorn, in fact, had a long family history as a colonial animal, introduced in Spain by North African invaders almost a thousand years before Spain in turn brought it to the Caribbean. "The Texas longhorn was of no undistinguished ancestry," but you wouldn't know it to look at one, according to an observer writing in 1878: "They are indeed nothing else than Spanish cattle, direct descendents of those unseemly, rough, lanky, long-horned animals reared for so long and in such large herds by the Moors on the plains of Andalusia."[13]

One of the chief virtues of Andalusian cattle was their adaptability, or described in the negative, their lack of "specialization." In Spain, cattle were not bred to favor one quality (like beef production) over another, resulting in animals that were neither fleshy nor particularly good for milk but nevertheless used for both.[14] Spanish ranchers maintained one breed of Andalusian cattle for bullfighting (feral descendants of which were also found in Texas) but otherwise allowed their cattle to determine their own evolutionary paths, which produced resourceful and independent, if not meaty, animals.[15] In sixteenth-century Mexico, this tendency to allow cattle to fend for themselves in reproductive matters was reinforced by a law against castrating bulls in order to ensure the productivity of the herds.[16]

One historian writing in the early twentieth century claimed that almost all improved breeds of cattle (that is, breeds developed with specific ends in mind) in Anglo-Saxon countries came from Great Britain because of the variety of breeds maintained there, these breeds' "superiority"—that is, their limited superiority in qualities valued by the breeders—and the "genius" of the breeders. Breeds from "pastoral countries" (implicitly non-Anglo, nonagricultural, and less advanced) by contrast were not improved and in some places hardly domesticated.[17] To a USDA employee in 1875, the cattle of Texas were of "mixed origin," "a medley of miscellaneously-bred animals, useful, in most respects, for beef, draught, and dairy purposes" but not at all "an improved breed." The animals grazing the uncultivated plains of Texas, New Mexico, California, and Colorado were, in fact, "comparatively valueless; nothing but the abundance of forage and the mildness of the climate makes them worth rearing at all." But these animals were not entirely without value. The "semi-barbarians of Texas and New Mexico" could be "worked up into well-conditioned [that is, fat and fleshy] beef animals when taken from their wild haunts and vagrant wanderings into closer pastures and under better care, by their more vigilant [Anglo] breeders."[18] When longhorns were eventually crossed with English stock, the longhorn heritage gave the offspring "self-reliance and initiative," and the Hereford or shorthorn stock gave it marketable weight.[19]

Herefords and shorthorns had both undergone improvement in the pastures of ambitious English breeders in the eighteenth century. Andalusian cattle might have had an illustrious pedigree in the remote Islamic past of Spain, but the names of the breeds and their breeders are not part of the longhorn legacy. English cattle, on the other hand, inherited a veritable royal lineage in association with their breeders. "Benjamin the Elder" was the first improver of Herefords in England and left his herd to his son, Benjamin the Younger, who continued to refine the breed as large beef animals.[20] According to one historian, T. L. Miller of Illinois became the "father of the Hereford tradition" in the United States when he imported Herefords in 1870,[21] but Henry Clay had imported Herefords as early as 1817.[22]

Shorthorn cattle could likewise trace their improved ancestry to a specific English herd (belonging to the Colling brothers estate), which was dispersed in 1810. These cattle were useful for beef as well as milk, but breeding to favor one quality generally reduced the animals' value for the other. Individual American shorthorns (and their pedigrees) were listed in a national register as early as 1846, and they were the first cattle crossed with longhorns in the effort to improve the latter in the 1860s.[23] Herefords were larger than shorthorns, though, and attracted attention at stock shows, including the centennial celebration in Philadelphia in 1876 and the first American Fat Stock Show in Chicago in 1878. Herefords seemed to be a better choice for western stockraisers. The demand for purebred Hereford bulls rose by 1880, when the American Hereford Cattle Breeders' Association founded its own "herd book," recording the pedigree of every Hereford that might be raised or sold as breeding stock.[24]

Perhaps not surprisingly, it was "cattle kings" that began the improvement of bovine commoners in the West, enriching themselves in the process. One of these was Charles Goodnight. He had tried breeding longhorns with shorthorns on his ranch in the Palo Duro Canyon without success since these purebreds did not adapt well to their southwestern environment. The best shorthorns, in any case, were sold in the East, where they were very popular, leaving only the "scrub" bulls for sale in the West. In 1882 Goodnight bought an entire herd of Herefords, each head worth on average four or five times as much as a longhorn, and was pleased with them. A longhorn steer might sell for $15, but Goodnight's Hereford crosses sold for $25 in the 1880s. By 1885, many of his hundred thousand cattle, grazing over a million acres, had the white faces characteristic of Herefords and their hybrid offspring.[25]

When the federal government wanted to establish two herds of purebred longhorns in 1927—not for beef, certainly, but for a wildlife refuge in Oklahoma and a game reserve in Nebraska—it found only a few dozen and decided to import bulls from Mexico to increase this endangered stock. Herefords had almost entirely taken over the western range. Not only were these animals more valuable than longhorns as beef, but they were docile and their appearance was uniform. "One of the most arresting changes in the West has been the appearance if its cattle," wrote journalist and onetime cowboy Paul Wellman in 1939. "No longer is the herd a mixture of colors, of long, curved horns, and of wildness. As alike in size and markings as a uniformed army it is now, and with a docility that removes it far from the adventurous trail herds of the old days. The true longhorn has almost disappeared. Here and there an occasional survivor can be found, kept as a curiosity. But the breed is more scarce in the West than the buffalo it supplanted."[26] "Now even the bulls are tame," wrote Mari Sandoz in 1958.[27] Improved breeding had domesticated and homogenized the "miscellaneously-bred" longhorn, rendering range cattle more dependent on human protection and more susceptible to disease but more completely exploitable than the "native" stock.

Sheep underwent a similar process of improvement. They were also of Spanish origin, their ancestors raised on the dry *mesetas* of Spain, some ultimately descended from African merinos. Shepherds had been raising the churro for millennia on the Iberian Peninsula, a sheep with long, coarse wool.[28] These were the sheep Columbus brought to North America. They did not adapt well to their first new home in the Indies but fared better in New Spain.[29] Merinos had been imported into Spain from Africa in the last quarter of the thirteenth century, launching a "Wool Revolution" in Europe as Spain shipped merino wool to European markets, which then fed the textile industries busy manufacturing cloth for a growing European population. Castillian merino wool largely replaced British wool in western European markets at the time, and Spanish merino investors and merchants became very wealthy from this trade, which added tax money to the royal coffers as well.[30]

Sheepraising in the Spanish colonies never became a lucrative business in part because wool produced there could not compete in European markets with wool produced on the Iberian Peninsula—indeed, with only a few exceptions, merino export was for a long time forbidden, and only churros were raised in North America[31]—whereas there was a growing market for American cattle hides. Also, many Spanish colonists came from primarily cattle-raising regions in Spain, and cattle demanded less protection and oversight in reproduction than sheep. Spanish stockmen therefore emphasized cattle over sheep in North America and did not import merinos until the eighteenth century.[32]

In Spain, breeders had maintained many varieties of merinos, raised mainly for their wool, but some historians claim these animals had never been formally improved.[33] Merinos were the sheep of privilege in Spain, held only by the clergy, nobles, and kings specifically for their high marketability;[34] it is unlikely that their owners would have neglected their care or breeding. It is possible, rather, that no "improvement" was necessary in an industry that had been at the forefront of European wool production for centuries. In the case of the lowly churro, neither Romans nor North Africans appear to have bothered to improve it. The advantage of these sheep to ordinary Spanish shepherds was their hardiness, which a breeding program to favor wool would have destroyed.[35]

To historians who value commercial qualities and the breeding programs that maximize them, the hardiness that churros maintained did not count as a kind of improvement. "History states," according to one sheep breeder in 1907, "that the native Spanish sheep was a very indifferent animal until improved by crossing with the Cotswold, which was imported into Spain from England in the eleventh century."[36] In any case, both churros and merinos were vital parts of Spanish sheepraising. Improvement recognized as such by breed historians began in earnest when merinos were imported into France and England in the eighteenth century, where they were bred to be more woolly, more meaty, or simply bigger. In France, merinos were bred for size and became the Rambouillet. The British bred merinos for the quality of their meat, and it was an improved British version, no longer a "common" Spanish sheep, that became commercially popular in the United States in the nineteenth century. In the United States, the merino was further improved for wool production, resulting in the delaine sheep, and bred for size, form, and quality of meat, producing the American merino.[37]

Spanish missionaries established flocks among many Native American groups in the Southwest, and wealthy Mexican landowners further stocked the southwestern range in the 1820s.[38] Northern European purebred rams were brought west with the gold rush and bred with ewes descended from this churro stock in California. Shepherds began trailing these crossed sheep north to Idaho in 1862, east into Arizona during a drought in California in the early 1860s, and into Montana in the 1870s. Tens of thousands of New Mexico sheep were trailed into Texas in the 1870s.[39] Sheep were trailed east from Oregon into Idaho and Montana as well, although these flocks were descended from Cotswold, Lincoln, and Anglo merino rather than Spanish stock. The number of sheep on the western range increased dramatically in the 1870s. Trails and pastures in Idaho, Montana, Wyoming, Nevada, Utah, and Colorado filled with hundreds of thousands of sheep, many of them Spanish churros or their descendants.

Whereas cattle can more or less fend for themselves as feral animals, sheep require more protection from predators, and therefore western flocks were not based on feral animals but sheep raised on Mexican ranches.[40] Even these sheep, like Mexican cattle, were expected "to hunt feed and water, to resist storms, to escape predatory pumas, coyotes, and bears, and in general to retain a preponderance of traits having little to do with carcass or fleece values." The churros' relative self-sufficiency compared with Anglo stock was the sign of their "degeneration," which "would be expected in any type of animal handled under similar conditions."[41]

Rambouillets had been imported since the 1840s (many of them reaching California after 1846), and by 1880 Rambouillet rams were in great demand for the improvement projects of western range sheep breeders. Like Herefords, which were introduced on other colonial rangelands, including those of Argentina, Uruguay, Brazil, Chile, Canada, Mexico, Australia, New Zealand, and South

Africa, Rambouillets were introduced in Argentina, Brazil, Russia, and Mexico, as well as becoming the most important breed of sheep in the United States.[42]

Although cattle and sheep competed for range forage in the 1870s and 1880s and both were potentially valuable stock for investors, cowboys held a legendary grudge against what some still call "meadow maggots." Paul Wellman suspected that some of this animosity had less to do with sheep and more to do with who the first large-scale sheepherders were: Mexicans and Native Americans.[43] According to Wellman, though, the sheep themselves were contemptible. He wrote, "Unlike cattle, sheep completely lack individuality, and their similarity makes them maddeningly monotonous." The sheepherders were allegedly dulled by the loneliness of their work and the insipidity of their charges, and to the cowboys they were "immeasurably inferior in the rudimentary social scale of the primitive West." Sheepherders were seen as "sullen" and "dreamy," and cowboys believed sheep "polluted the country in such a way that cattle would have nothing to do with it."[44]

But if there was money to be made, sheep would do. In 1869 twice as many sheep as cattle were grazing in Colorado Territory, most of them between Denver and Wyoming.[45] By 1881—near the height of the cattle bonanza—more sheep than cattle were raised in Montana.[46] Cattlemen who faced losses in depressed beef markets, bad winters, or bovine diseases had been turning to sheep since the 1870s.[47] The real boom in sheepraising, though, came between 1890 and 1910, when the cattle industry was severely hit by losses. Cattle and sheep alike were then relegated to fenced pastures and subject to the intensifying methods and purposes of control on the part of their keepers. Many large cattle operations turned to raising sheep in that period. In Wyoming, for example, the number of cattle declined after 1887, whereas the sheep population—less than half the number of cattle in 1887—grew steadily, even precipitously, in the early 1900s.[48]

By the turn of the century, western stockraisers were more concerned with the internal workings of their animals than they had been during the period of the open range, when animals had been valuable primarily in number only. The profitable reproduction and improvement of range animals demanded attention to the quality of individual male animals, the breeding "efficiency" of individual females, and reproductive behavior.

With respect to cattle, herd improvement began with the replacement of longhorn or otherwise "scrub" bulls with purebred Herefords or shorthorns and the castration of bull calves. Ideally, the bulls were to be kept separate from the cows until the cows came in heat, at which time the bulls were allowed to "service" them. The relatively high price of bulls encouraged ranchers to breed them with as many cows as possible, but a sensible rancher limited the number to no more than twenty-five or thirty cows for each bull he turned out. This way, "the bull is strong, vigorous, and fertile." A rancher scheduled the intercourse of his cattle so that calves would be born as early in the spring as possible, producing a uniformly well-developed calf crop to be weaned in the fall.

Although the breed of the bull was extremely important, the cows might be any breed at all. A range cow was an unmistakable embodiment of feminized nature—the ground out of which improvement was made. Her offspring became both the sign of herd productivity and the evidence of better breeding. "No bull is too good to use on scrub cows," wrote one range expert in 1917. A rancher was foolish not to eliminate "non-breeding or shy-breeding cows," he commented. Left to themselves, cows had a breeding "efficiency" of about 60 percent on the open range, but "carefully managed," this figure could be improved to 85–90 percent.[49]

For sheep as for cattle, the breeding flock was primarily female, and rams carried the signification of the breed that was brought to fruition in the bodies of ewes. Prolificacy (multiple births) was a quality to be desired in a ewe. Twins in sheep are common, and when encouraged by a careful breeder, the weight of lambs produced per ewe might be systematically increased.[50] A good ram was understood to be the best way to improve a flock, and his "vitality," like a bull's, was an object of concern to the breeder. The ram's vitality should be conserved by "tagging" the ewes (clipping the wool under the tail around the vulva) and allowing the ram to "serve" each ewe only once rather than many times.[51]

The overall project of animal improvement emphasized the marketable qualities of domesticated animals. This was especially clear as one Indian agent after another tinkered with the stock of the Navajos from the 1880s through the 1930s. These were the oldest variety of American sheep, "descended from the low quality strain introduced by the Spaniard," whose wool was "long, coarse, and practically worthless as an article of commerce."[52] The Navajos' original flocks produced lambs in all seasons, so there was no "crop" as produced under Anglo flock management. The sheep were also generally hardy, but the flocks did not increase much in number due in part to winter losses. Kit Carson destroyed most of the original Navajo sheep stock in his campaign against the Navajos in 1863 and 1864. After the Navajos had been removed from their reservation and then returned to it in 1869, they were given 14,000 new sheep (probably of Spanish or even Navajo origin) to begin their pastoral life anew, receiving 10,000 more in 1872.

Agent D. M. Riordan began "improving" the flocks in 1882 when he introduced merinos from California to increase the yield, length, and quality of fleece and encouraged the Navajos to eat their culled ewes (eating a genetic archive in the process). Fine-textured merino wool was best suited to the industrial production of textiles for which it had been developed in Britain; it was not at all suitable for the hand spinning and weaving done by Navajo women. A different agent, J. H. Bowman, began asking for Cotswolds (longer-wooled sheep) in 1884 to solve the problem but never received any. Successive agents took up the breeding project and dropped it again when they were transferred elsewhere, leaving a "multiplicity of sheep types introduced without any organized plan," which resulted in "a heterogeneous and very inferior sheep." Shropshires were brought in after 1900 to improve the Navajo sheep as mutton, but these animals destroyed the flocking instinct of the Navajo sheep and demanded increased herding. The animals were then crossed with fine-wooled rams (like merinos) to restore flocking, but the wool staple shortened too much and the fleeces became too greasy. By 1934, Navajo rugs, which had previously been of unparalleled quality and beauty, were "a mass of knots, unevenly scoured and unevenly dyed" and the quality was declining every year, according to one observer.[53]

Another generation of experts at that time set aside a flock of Navajo sheep purchased from areas of the reservation where crossbreeding had not been common and bred this flock to restore fleece suitable for the hand weavers as well as for commercial wool sale and to encourage greater body weight and early maturity. One breeding program repeatedly mated the native stock—a process known as line breeding—crossing individual animals with desirable characteristics primarily to improve wool quality for hand weaving. Another breeding program crossed native ewes with Corriedale and Romney rams, whose wool could be used in hand weaving as well as mechanized weaving and whose greater body weight and early maturity might enhance these qualities in the native stock.[54]

The sheep-breeding program was part of Commissioner of Indian Affairs John Collier's Indian New Deal and involved the establishment of a sheep-breeding laboratory at Fort Wingate through the cooperation of the Indian Service and the USDA Bureau of Animal Industry. Unlike other Indian Service programs imposed on the Navajos, this one was not controversial, but its results were mixed. The researchers were most successful in developing a native sheep with uniform fleece color and texture; they were less successful with the Romney and Corriedale crosses, whose wool was too fine for hand weaving. Both parts of the breeding program yielded heavier fleeces, although the weaning weights of lambs (a function of the rate of maturity) had not increased much. Other crosses were attempted after 1942 to specifically address this problem.

Meanwhile, mutton-type Rambouillets had been purchased in great numbers by another federal agency on the reservation known as the Navajo Service, made up of officials from the Indian Service, the Soil Conservation Service, and other New Deal emergency programs. Here, indeed, was the double edge of imperialist nostalgia, as one part of the federal apparatus facilitated Navajo sheepraising as a stock industry similar to Anglo stock industries, while another part attempted to restore to the Navajo what had been destroyed by those very processes of colonization and commercialization that had produced Navajo immiserization in the first place. Navajo herders themselves preferred the Rambouillets to the Fort Wingate sheep, and sheepraising shifted away from production for hand weaving to the sale of lambs for slaughter. Had Navajo herders been systematically included in the process of designing sheep from the beginning, the program of research (and the sheep themselves) would probably have looked quite different. By 1948, only 10 percent of Navajo wool was used for rug weaving; research at Fort Wingate focused on fine-wooled sheep for the commercial wool market, a focus it maintained until it closed in 1966.[55]

After so much introduced degeneration in the name of "improvement" and so little collaboration with the people whose sheep drew the attention of several federal agencies, the Navajo sheep allegedly faced "a more satisfactory future" in 1948, according to one contemporary assessment.[56] This story serves as an apt cautionary tale as the USDA continues to exert genetic control over other people's plants and animals in other parts of the world in the name of "improving" their commercial qualities. In the case of the Navajo sheep, the entire process of "improvement" ultimately depended on the "old type" of sheep for any success whatsoever. This fact was surely not lost upon the shepherds.

Scientific stock breeding promoted qualities that had comparatively little to do with meat or wool as marketable commodities. In the 1940s, the good breeding cow was quite a piece of work, culturally speaking. She had "a straight top and bottom line, a broad back, deep wide hind quarters and a smooth, heavily muscled shoulder. With these predominate beef points the cow should also possess a short, well-molded head on a short, neat neck, trim legs, and an even coating of silky, glossy, fine hair. She should also possess a large, well formed udder to insure ample milk for her calf, and a pleasant disposition evidenced by a docile eye and a willingness to be handled."[57] A polled shorthorn cow whose photograph appeared in a 1940 livestock management text was described as "very feminine." The author continued, "Her even lines, beautiful head, and thickness of fleshing makes her a splendid example of brood-cow character."[58]

An Aberdeen Angus bull pictured in an animal husbandry handbook in 1949 was described by the author as "handsome," and the reader was directed to note "the masculinity of head and neck, the massive body on short legs and the straight top and bottom lines." Masculinity was "essential," but "coarseness in the herd sire is anything but desirable."[59] "Coarse" is apparently an aesthetic judgment and seems to apply to an animal's standing appearance as well as to the fluidity of its gait. A bull should have a short, wide head with "well-distended" nostrils; a large muzzle; eyes wide open and clear; a deep, wide chest; deep "well-sprung" ribs; a "roomy middle"; activity on foot; and a "rugged bone." Whereas bulls were expected to exhibit ruggedness and massivity, good cows were distinguished by their "refinement" as well as by their docility.[60]

Sheep were generally described in the same terms, although their literal blockiness (depth, width, and straightness of lines) appears to have dominated the categories in which they were judged.[61] A good ram had "a deep voice, strong head and neck, and bold carriage"; he had a "fairly wide, masculine head that is strong but not too coarse." Large ewes were generally better than small ones, "if not too coarse," and a good ewe "is refined about the head and neck, and has a roomy middle, a broad rump, and good udder development." Ewes were also expected to have a "quiet, motherly disposition" to ensure quality care of their offspring.[62] Ewes "well along in years" (as long as their mouths are not "broken" or missing teeth) could make valuable breeders, "for no other animal is liable to give less trouble or better returns to the novice than these," and the fact that they were "staid and matronly is an assurance that they are of more than ordinary worth as mothers."[63] One breeder gave the Shropshire ram special praise as an animal whose "compactness of form is ever liable to deceive us as to his correct weight, and whose masculine character and mutton qualities stand out at all points of his anatomy, so much so that we cannot fail to recognize in him something of an Adonis and a Hercules in the animal kingdom."[64]

The vocabulary of animal gender in part reflects the perceived importance of certain secondary sex characteristics in animals kept specifically for their ability to reproduce. But this vocabulary is also clearly saturated with a human gender ideology. The beauty and docility of a trim-legged cow, the quiet disposition of the matronly ewe, the handsome ruggedness of a deep-chested bull, the deep voice of the ram, together with the class implications of "coarseness" and "refinement," suggest that a major component of breed improvement was not merely an increase in the quantity of marketable goods produced by a given animal but the projection of social hierarchies of class and gender onto the bodies of domesticated animals. The breeds themselves reflected a cultural preoccupation with race and descent, the best varieties (not surprisingly) thought to be found in northern European and Anglo-American pastures. The social construction of "good breeding," identifying people of socially superior family and education, may appear to be derived from the idea of animal breeding (meaning improvement—breeding *up*). But the concept is perhaps best understood as social and specific to a society in which cultivation in all matters, from agriculture and animal husbandry to the arts and sciences, is the sign of civilization. An improved animal of a given breed reflects the "breeding" of the society that created it.

Twentieth-century animal husbandry, supported by research in agricultural colleges and experiment stations, predictably found ways to manage the reproduction of cattle and sheep ever more precisely. By the 1930s, scientists in animal husbandry developed techniques of artificial insemination, whereby the sperm of a particularly valuable individual could be used to impregnate many more females than was possible under ordinary circumstances. But the most sophisticated reproductive technology has always been more widely used in dairy herds, where the animals are kept in closer quarters under greater supervision than range animals.[65]

While range stockraisers were busy improving their herds and flocks, they became increasingly aware of the nutritional needs of their animals as well. Animal nutrition demanded a knowledge of the composition of plant substances and how food is digested and used in the animal, not merely the amounts of food and water an animal needs to thrive. A precise and scientific regulation of animal feeding was necessary to guarantee the efficiency of animal digestion—that is, the efficiency with which an animal converts its feed into materials useful to its feeders.

A popular handbook on animal feeding noted that "the animals of a farm should be regarded as living factories that are continuously converting their feed into products useful to man."[66] Arthur Sampson, prominent in the development of interdisciplinary range management research, echoed this popular view. In 1923 he wrote that animals are "living factories" that run on "vegetational fuel" and are "continuously manufacturing flesh, leather, wool, mohair, motive power, and numerous other valuable products for the benefit of man."[67] Another scientist wrote in 1917, "The essential function of the animal in a permanent system of agriculture is the conversion of as large a proportion as possible of . . . inedible products [solar energy, protein stored in grass and crops] into forms whose matter and energy can be utilized by the human body." The efficiency of this conversion required "as intimate a knowledge as possible of the fundamental laws governing the nutrition of farm animals," which had virtually nothing to do with them as living creatures and everything to do with the "physico-chemical process" of storing energy in the animal body that could be released and used by people.[68] Of course, this view of nutrition was not reserved for animals. The science of human nutrition developed along the same lines at the same time, beginning in the 1880s, and remains with us as we count calories, vitamins, minerals, cholesterol, protein, and fat in a mechanistically reductive process of calculating the "value" of our food.[69]

The farm or feedlot was the ideal place to practice the most precise feeding regimes, just as the dairy was the best place to practice artificial insemination. But although range animals were generally responsible for feeding themselves, virtually every other aspect of their grazing environment came under the managing hand of stockmen and range scientists. Range forage became a crop whose composition, growth, and harvest could be controlled for the maximum yield of both forage material and the animal products into which it was converted. This transformation of the range into an agricultural entity had many expressions, including the direct harvest of grass and forage plants for hay as well as the improvement of pastures by irrigation, the replacement of wild with domesticated grasses, and the regulation of animals' grazing patterns and schedules to encourage the productivity of the range as well as of the animals themselves.

The science that was developed to address these complex relationships between animals and the range was itself complex, including animal husbandry, veterinary science, botany, agronomy, and soil science, among others. Range science emerged in the 1920s as an interdisciplinary field profoundly informed by an ecological point of view, focusing on the relationships between grazing animals, the range that supported them, and people's demands for the productivity of both. This science was developed specifically for the western range. It made possible the theoretical transformation of grasslands into agricultural places, complementing the improvement of commercial pastoralism begun by animal breeders and hay mowers.

The most direct agriculturalization of range plants (and perhaps the first attempt at western range management) involved keeping the stock off parts of the range and cutting the grass for hay. Few ranchers put up hay before the 1880s. In the North, where winters were severe, cattle were subjected every winter to "nothing less than slow starvation; a test of stored flesh and vitality against the hard storms until grass comes again."[70] Norman Colman, commissioner of agriculture in 1887, wrote that the "wastes of the past have been enormous," among them the irregular feeding and semistarvation of range animals.[71] Moreover, eastern purebreds were less likely to "rustle" for their feed than longhorn stock, increasing their risk of starvation if no hay was forthcoming from their keepers. Winter losses as high as 10 percent were not regarded as extreme. During the disastrous winter of 1886–87, losses ran as high as 90 percent.[72] In the South, where winters were more mild, drought could present as serious a threat on overstocked and depleted rangelands. One ranch near Fort Worth lost 15,000 out of 25,000 cattle during a drought in 1882–83.[73] Bad weather and range conditions combined with the large numbers of western livestock brought panic to the stock market in the 1880s, and it collapsed decisively in 1887. Animals had sold for $50 or more apiece in 1882 at the height of the boom. In 1887 they were unloaded by speculators for as little as $10 a head, if they could be sold at all.[74]

Ranchers were already beginning to invest in leases, land purchases, enclosures, and herd improvements by then, and the collapse of 1887 further encouraged them to secure their investments. Providing winter feed was one way to minimize losses. In 1880 only about 80,000 acres of grassland were cut for hay in Montana and Wyoming, but by 1900 this figure had increased more than tenfold, to over a million acres.[75] Haying in summer and feeding in winter came to be part of the cowboy's seasonal round on western ranches. By 1900, fences kept stock off the haymeadows as well as out of neighbors' pastures.

One source of hay was the grass simply found, not planted, on the range. Its productivity could, of course, be improved by irrigation where the grass was valuable enough and where water was close by. Floodwater in rivers could be turned out onto the range through canals once a year, "improv[ing] the natural grass crop in favorable areas."[76] More than half the hay crop consisted of "wild" irrigated grasses in Wyoming at the turn of the century.[77] In 1905 USDA researchers reported that irrigated western wheatgrass (a native species found primarily west of the Dakotas) made such good hay that it was bringing better prices than timothy, a popular domesticated grass. Indeed, western wheatgrass appeared valuable enough at that time to justify its domestication so that it could be distributed and planted to improve less productive meadows.[78]

But wild hays were generally less valuable than tame grass hays and were preserved only on land where nothing else could be sown. In general, an irrigated pasture was considered profitable "if it yields returns equal to those from an alfalfa hay crop on similar land," indicating that tame forage crops became the standard by which hay and pasture productivity was measured.[79] The USDA had recommended replacing midwestern prairie with tame grasses as early as 1862: "A good range is one of thousands of acres of high rolling prairie through which runs a never-failing 'branch,' whose banks are dotted with small groves. A *better* range is the same territory with the prairie grasses killed out, and blue grass in their stead."[80]

By 1924, slender wheatgrass was the only native grass whose cultivation had been found profitable, according to the agrostologist in charge of USDA forage-crop research at that time.[81] The same attitude prevailed in 1948: "Unimproved native pastures on the whole have a lower carrying capacity than improved tame pastures. Some of our most productive tame pastures will carry two cows or steers on an acre for a grazing season 8 months long. Native grazing lands that will sustain a cow on less than 8 acres is rare."[82] Tame hay could be further improved by irrigation where it was practical. About 88 percent of tame hay pastures in Wyoming were irrigated as early as 1899.[83]

As the range livestock industry got down to the business of improving its herds, flocks, and pastures, it was the work of the Division of Botany, established within the USDA in 1868, to identify and study the growth habits of range plants, including grasses. Botanists' work intensified in the 1880s, when the depletion of the range was everywhere apparent, and the Division of Botany began investigating the possibility of introducing domesticated grasses and forage plants to improve the range in 1888. This work was explicitly intended to address western range problems and was begun at the only experiment station at that time conducting research on grasses, at Garden City, Kansas.[84]

Range research was a subfield of botany but was considered important enough in 1894 to justify the establishment of a Division of Agrostology within the USDA, devoted entirely to the study of range plants. Under the direction of F. Lamson Scribner (who botanically identified and named western wheatgrass), the new division was to prepare a report on grasses and forage plants. Agrostologists were not only concerned with studying or propagating range plants close at hand—like slender and western wheatgrass—but were interested in gathering new varieties from abroad as well. One of the main objectives of the first federally sponsored exploration of foreign plants, in 1897, was to identify and procure hardy, drought-resistant strains of alfalfa for introduction in the West. The most promising varieties came from Turkey and Russia.[85]

In 1901 both the botany and the agrostology divisions were absorbed into the Bureau of Plant Industry, and by 1910 the bureau was promoting the cultivation of clovers and alfalfa for hay in the West as well as seeding mountain rangeland with timothy, redtop, or Kentucky bluegrass, depending on the acidity of the soil.[86] Range researchers continued to recommend cultivated, domesticated grasses and forage plants for hay, and for pasture where possible, through the 1940s. "Desirable species," like alsike and red clover, increased the productivity of wild hay meadows, and meadows "plowed and reseeded to mixtures such as timothy and red clover or smooth brome, and alsike clover, yield almost twice as much as before seeding" in many western states.[87]

Other species introduced in the West included crested wheatgrass, a non-sod-forming, drought-resistant perennial bunchgrass native to Turkestan, which was introduced in 1898 and became valuable in reclaiming abandoned wheat land in North Dakota in the 1920s.[88] The sod-forming perennial intermediate wheatgrass was introduced from the Soviet Union in the 1940s for use in the northern and central Great Plains and the Pacific Northwest.[89] On the southern Great Plains, where Dust Bowl erosion was worst, both native and introduced grasses were used to stabilize soil as quickly as possible. One of these was weeping lovegrass, a perennial bunchgrass introduced from Tanganyika into South Africa in 1927 and from there imported for conservation purposes in 1934 in the Southwest.[90]

Alfalfa was an especially significant forage crop in the West, with a long history as a crop of imperial conquerors. Named by the Spanish from its designation in Arabic, which means "the best fodder," alfalfa is the oldest crop cultivated only for forage. It came from Media, and the name of its origin was preserved in the Latin name of the *Medicago* genus. The Medians conquered Persia in the eighth century B.C. and were conquered in turn by the Persians two centuries later. Then the Persians introduced alfalfa in Greece in the fifth century B.C. The forage invader outlasted Persian military occupation. Romans adopted alfalfa apparently after conquering Greece, and from there, alfalfa spread west and north and arrived in North America with the Spanish. Whether alfalfa came to Spain via Rome or North Africa is not clear, but the survival of an Arabic name for the plant in Spanish suggests the latter. "Thus the queen of forage plants," wrote a forage-crop expert in 1950, "followed the path of historic civilizations and conquering armies from East to West."[91] It would seem that a colonial enterprise was hardly complete without it.

This is not surprising. The planting of forage crops is consistent with the practice of colonialism as the maximization of certain parts of agricultural production at the expense of other parts, especially the production of food for people. Unlike other forage crops that could be eaten by people as well as animals, alfalfa was strictly cultivated for animals and provided a richer diet for them than just grass. Its cultivation required that productive land be set aside for animal, not human, food production. Animal husbandry in this case was not merely a way to put inarable land to use but involved the improvement of pastures—replacing grass with other plants in an animal's diet either to increase milk or meat production or to more closely manage the grazing area and habits of animals. Alfalfa culture tended to take land out of local food production and intensify animal production, probably for the benefit of the powerful rather than those from whom the land was taken, and thus provided a profitable colonial tool for conquerors.

Whether that was the case in Persia, or Greece, or Spain is not clear. We may be sure, however, that armed invasions did not simply offer benign cultural "exchanges" to the subjugated people and that changes in agricultural practices as a result of invasions were probably associated with the external imposition of new rules for distributing land and food. In any case, alfalfa was certainly such a colonial tool in the West, where it bolstered a large and growing livestock industry, securing arable land to feed animals that would in turn enter the economy like wheat—ultimately edible but primarily remunerative.

In the 1920s, alfalfa was "the second most important forage crop in America, being exceeded only by timothy," and under irrigation in the West, it was the most productive forage crop. The director of USDA forage-crop research in 1924 wrote of alfalfa, "The agricultural development of western America is to a large degree associated with the culture of this plant."[92] Alfalfa alone occupied nearly half of the western acreage devoted to forage crops between the Rocky Mountains and the Pacific coast in 1948.[93]

Although there was a clear distinction between a hay meadow planted in alfalfa and a hay meadow left in "wild" grasses, the wildness of the wild hay is an ambiguous matter, similar to the ambiguity of the longhorn as a "native" bovine. Red clover, alsike clover, redtop, meadow fescue, and timothy had all been introduced from Europe and escaped cultivation long before they were cataloged and studied as part of the western range or reintroduced in an effort to improve range condition and productivity.[94] Timothy had become naturalized so successfully that some range investigators believed it was indigenous. It had never been cultivated before it was brought to North America, and its large-scale deliberate introduction here probably explains the extent of its naturalization.[95]

Smooth brome was introduced in California from Hungary in 1884 as a hay and forage crop and likewise escaped, gaining "sparse and scattered footholds on many of the mountain ranges" by 1934.[96] Even downy brome, which has become the scourge of intermountain ranges, was introduced deliberately at least once to "improve" the range.[97]

There is every reason to believe that "wild" hays have long included once-domesticated varieties. On the other hand, these species were generally recognized as superior to indigenous species (downy brome notwithstanding). Red clover especially was considered a mark of range pasture quality since it was known to be palatable to livestock and was associated with great historical significance: "Red clover has contributed even more to the progress of agriculture than the potato itself, and has had no inconsiderable influence on European civilization. Its cultivation has led to an increased production of stock as food for man and in this way has fostered and advanced commerce, industry, and science."[98] Although at least eighty species of clover are indigenous to the United States, none of them "have proved to be of agricultural value in this country, although they contribute to grazing and to the wild hay crop."[99]

Naturalized European grasses and clovers underwent the same kind of improvement to which the longhorns had been subjected as researchers from the early 1900s through the 1940s developed varieties of domesticated grasses for reseeding specific western areas. The seeds of indigenous grasses were harvested to improve worn out or eroded pastures and hay meadows and were also of interest to researchers in the 1940s as they continued to determine "what native species justify selection for improvement."[100]

Although a great deal of range research in the twentieth century was remedial in nature—focusing on how to restore overgrazed ranges—what was restored was range *productivity*, an agricultural quality that could always be maximized, whether the range was actually replanted or not. Improvement might take the form of plowing and reseeding (or simply overseeding) a hay meadow or pasture with desirable species that were then harvested more or less like other crops, but as early as 1920, artificial reseeding of depleted ranges for remedial purposes was not generally recommended.[101] The improvement of most areas that were too steep, too dry, too remote, or too fragile to justify direct reseeding took the form of managing livestock to promote the plants' own capacity for regeneration.

The difficulty of this type of range management lay in determining the condition and carrying capacity of the range, issues that had occupied the Forest Service almost from its establishment. In the days of the livestock boom, the condition of the livestock indicated in a general way both range condition and its carrying capacity—logically enough, since a depleted or overstocked range produced thin and unhealthy animals.[102] Some experienced and thoughtful stockmen were as concerned about range condition as they were about the condition of their stock, and they understood range condition and carrying capacity in more complicated terms.

Albert Potter, an Arizona sheep rancher, was one such stockman. He traveled to Washington, D.C., in 1899 as a representative of the Arizona Wool Growers Association to protest the prohibition of grazing on the new forest reserves. Potter so impressed Division of Forestry chief Gifford Pinchot with his knowledge of forage plants and range condition as well as livestock management that he was hired to develop grazing policies for the division in 1901. Potter was appointed head of the Grazing Branch of the new Forest Service in 1907 and remained in government service until his retirement in 1920. His outline for grazing regulations, published in 1908, took the location of water, topography, soil type, the relative value of available forage, the amount of pasture required to sustain an animal on different range types, and the kind of animal itself into consideration in determining how intensively a given range should be used.[103]

It was this point of view that informed the development of range science in the twentieth century, beginning with Potter's tenure and the establishment of the Forest Service. Range research collected data to support the kind of recommendations made by Potter, and in the process, range forage became a crop whose productivity could be measured and managed, without ever plowing, planting, or cutting. The uncultivated range, if harvested properly by livestock, reproduced itself and maintained its own productivity. The optimum relationship between range and livestock productivity was a minutely local matter, dependent on the plant species available, their growth and reproductive requirements, soil and water conditions, climate, and the type of livestock grazing. Among the first range experiments undertaken by the Forest Service, before 1910, were studies of deferred and rotational grazing in the Wallowa National Forest in Oregon. If stock could be kept off the range until plants had gone to seed, grazing would actually facilitate natural reseeding and germination. In the case of rotational grazing, one range section out of five or more would remain unused each year. Yearlong grazing that took no account of the growth and reproductive requirements of the grass was the most destructive to range forage crops. Protecting grass from grazing increased the productivity of living plants but did nothing to encourage the growth of seedlings.

Arthur Sampson, one of the early range researchers and the first to produce a range management textbook (in 1923), consequently recommended deferring grazing until the grass had gone to seed.[104] The amount of available forage palatable to livestock as well as the amount of forage necessary to maintain an animal were other factors that affected how many animals could graze a given area without interfering with the range plants' ability to maintain and reproduce themselves. Sampson and his colleague James T. Jardine had begun relating forage types and carrying capacity in Oregon. After Jardine was chosen in 1910 to direct a new Office of Grazing Studies (again in the Forest Service), he established a systematic research program—the range reconnaissance—to be implemented on federal forestlands based on the kind of work he and Sampson had been doing in Oregon. Range reconnaissance teams collected data on the amount, type, and degree of utilization of vegetation, topography, acreage, range condition, location of water, and other information, all of which would be used to make a range survey map and develop a grazing management plan for a given area.

Unlike the estimates of range condition and carrying capacity made in the first decade of the national forests, even the most careful of which were done by "rule of thumb,"[105] range reconnaissance results were intended to be scientifically accurate and demanded a higher degree of botanical expertise, not just a general familiarity with the range or stock management. Over time, the carrying capacity of a given area could be determined, and since range surveyors carefully recorded the data, carrying capacities could be recommended for ranges of similar type.[106] Botany students became the surveyors of choice, examining the land meticulously (and at considerable expense) section by section under the supervision of experienced range scientists.[107] Even as late as 1935, most range positions were still filled by men trained in the disciplines that made up range science, including botany.[108]

Sampson recognized the need for specially trained range officials, and in 1919 he outlined a grazing curriculum to be offered in forestry schools. Students would take courses in zoology, entomology, chemistry, and animal nutrition as well as botany; they would study native, cultivated, and poisonous range plants; and they would learn methods of improving and managing forage and principles of stock management, fencing, and range economics.[109] The complex and interdisciplinary nature of range science advocated by Sampson was the result of his own training. Sampson had been a student of ecologist Frederic Clements at the University of Nebraska. Ecological principles that described communities of different organisms and their changes over time profoundly informed range science as it developed in the Forest Service.

Clements's theory of plant succession, based on research in the grasslands, provided the theoretical groundwork for studies of range deterioration or improvement in relation to the intensity of grazing allowed in a given area. According to Clements, a plant community developed through successive stages like an organism until it arrived at a climax stage, characterized as "adult," at which it was capable of reproducing itself.

Clements himself used what he called the "succession of races and cultures" around the world, including American western settlement, as an analogy for plant succession. He cited waves of invasion in ancient Europe and Mesopotamia and then turned to a history "better known to us: The series of invasions that have swept over England, involving Pict, Goidel, Brython, Roman, Angle and Saxon, Dane and Norman. A similar succession on our own continent is illustrated by the Maya, Toltec, Aztec and the Spaniard in Mexico, by the various Pueblan cultures of the Southwest, and by the trapper, hunter, pioneer, homesteader, and urbanite in the Middle West." (This homology between natural and social historiography has been noted by other scholars, most recently Donald Worster and Stephen Pyne.) Pioneer, ruderal species prepared the ground for further waves of plants on the grasslands until perennial native grasses became established in a stable climax. Disturbances, like fire, drought, or grazing, could disrupt a stable climax community or arrest the progress of communities lower on the scale.[110]

Arthur Sampson, who along with James T. Jardine lay the ecological foundation of range science before 1920, wrote that the management of the "forage crop" demanded that the changes in range plant composition be recorded from year to year to determine how well "forage species are holding their place against grazing."[111] Range management became "the art and science of procuring maximum sustained use of the forage crop without jeopardy to other resources or uses of the range."[112]

The most productive range was one characterized by perennial grasses, which constituted the "climax herbaceous" stage of range plant succession. Climax vegetation provided the best soil protection, the greatest variety and quantity of forage, the highest grazing capacity, and the "greatest production of livestock products."[113] Overgrazing destabilized the climax grass community, allowing it to degenerate into less productive stages by killing out perennial grasses and allowing poorer forage plants (annual grasses and perennial and annual weeds) and even poisonous plants to predominate. The density of plant cover—even the density of grass cover—did not determine range condition, but the actual identities of the plants growing in a given area (and their relative palatability), combined with the "progression" or "retrogression" of these communities with respect to the climax stage, determined the condition of the range. Revegetation after overgrazing could be induced by "cropping [that is, either grazing or cutting] the herbage in such a manner as to interfere as little as possible with the life history and growth requirements peculiar to the different successional plant states."[114]

Since many factors might interfere with the establishment of a "natural" climax, the point at which progressive revegetation stopped—the subclimax, characterized by perennial grasses like wheatgrass, whether native or not—became the stage that represented the best range condition. Optimum grazing allowed the subclimax grass community to maintain itself indefinitely without degeneration. An area's location on the scale of vegetative development, as determined by the careful measurements of range reconnaissance, served as an index of its carrying capacity and promised a reliable standard for regulating grazing on forest property.

Range science as practiced by Sampson and his colleagues was both experimental and applied, but it took place exclusively within the domain of the Forest Service from the 1890s to the 1930s, and relatively little extension work was done in that period to transfer range scientists' knowledge/power to stockmen themselves. Range research was confined to a few types of western range in national forests for decades, and research results were often so locally specific that general dissemination would not have been useful.[115] The Forest Service had no range extension program at all until 1923, and even then education outside the Forest Service was not a priority.[116] The theoretical agriculturalization of the range was used primarily to educate range officials themselves and to inform state regulation of forestlands, not to change the nature of grazing practices in general.

A major factor limiting the influence of the Forest Service in matters of range science education, in contrast to its influence in spreading the gospel of forestry, was the fact that most rangeland on federal as well as state and private lands lay outside the jurisdiction of the national forests. The Forest Service administered a much greater percentage of western timberland than grazing range. Whereas at least 50 percent of all western commercial timberland was under Forest Service administration, the Forest Service managed less than half of federal rangeland and only 14 percent of the western range as a whole.[117] Grazing studies and range science were important components of forest administration, but they were only components. Even the acceleration of research in the 1920s, as a result of the coordination of grazing studies with other forest research in 1926 and new appropriations under the McSweeney-McNary Forest Research Act in 1928, was not predicated on range management extension work but on establishing more firmly the principles for regulating livestock use of national forest ranges.[118]

Other federal priorities got in the way of even the Forest Service's regulatory policy. During World War I, the Forest Service was directed to permit the "fullest utilization" of its range property, resulting in a significant increase in the numbers of cattle and sheep grazing in the national forests.[119] The greater numbers of stock were justified by expected increases in wool and meat production for the war effort, but these never materialized. Stock came out of the forests thin and weak; the range was predictably depleted. After the war, Albert Potter directed forest officials to return to enforcing conservative grazing regulations, but a new war effort a few decades later would make the same demands on federal ranges.

In 1944 range scientists were confident that thirty-five years of research would "greatly improve the prospects for sustained production of meat, hides, and wool needed by the United States in World War II," but they already foresaw that changed "biological, economic, and social conditions will present pressing range-production problems" after the war ended.[120] By the 1940s, range science was expected to support maximum range production of animal materials rather than forage during wartime, while hopefully preventing the worst damage of overstocking.

There is no question that range science dramatically multiplied the number of objects brought under the experimental scrutiny and managerial control of Forest Service range experts and that these experts understood forage as a crop. It is also true, however, that range regulation was (and still is) enormously unpopular among stockmen and the politicians who spoke for them and that there was no consensus inside or outside the Forest Service that the experts could actually deliver the increases in range productivity that their complex science promised.

Even by the Forest Service's own admission in 1936, less than half of the range on national forestland was in "reasonably good condition."[121] The Forest Service claimed to have restored forestlands to 70 percent of their potential productivity and to have increased the productivity of the range 19 percent between 1910 and 1936.[122] But given the complexity and slow progress of range reconnaissance work, and the fact that this work did not even begin until 1911, it is not clear how the Forest Service arrived at these figures. Walt Dutton, a forester from the Pacific Northwest, agreed in 1975 that the condition of the public domain in the 1930s was generally poor but did not see national forest range, for all its protective administration at that time, as an exception.[123]

The rejection of Forest Service range expertise and regulation was apparent when the authority to regulate public range outside the national forests was given to a new bureau separate from the USDA instead of to the Forest Service. Colorado congressman Edward Taylor proposed regulating grazing on all public lands, and the resulting 1934 Taylor Grazing Act created a Grazing Service in the Department of Interior to organize and administrate grazing districts. The needs and interests of stockmen who sat on the administrative boards of the new grazing districts, rather than range management principles alone, determined the regulations. The publication of *The Western Range* by the Forest Service in 1936 was an unsuccessful attempt to show that it was the proper agency to manage all of the public range. The Forest Service was eventually placed in charge of what had been drought-stricken and eroded lands purchased in the 1930s under the National Industrial Act and the Emergency Appropriations Act, but this transfer did not take place until 1953.[124] Until that time, the Soil Conservation Service administered programs to stabilize and restore productivity to Dust Bowl lands.

Unlikely to create a revolution in grazing practices on public lands, the Grazing Service already represented too much government involvement in western stockraising to satisfy many stockmen and the politicians representing them.

The Grazing Service used some range management principles, particularly the animal-unit-month (AUM), based on surveys similar to those undertaken on national forest rangelands, in calculating how many animals should be allowed to graze an area. An AUM is how much forage it takes on a given range type to support a cow and calf, or five ewes and their lambs, for a month. It is a calculation that theoretically takes into account the relative palatability and density of all available forage and maximizes the production of animal if not plant material.[125] For all of its apparent standardization, however, the calculation of AUMs varied from agency to agency, an inconsistency that discredited the range survey as a reliable knowledge-gathering method.[126]

Regulation of any type, especially given the unreliability of regulatory standards, together with the imposition or increase of grazing fees, was a central object of Nevada senator Patrick McCarran's campaign against the Grazing Service, which began in 1941 and extended to the General Land Office and finally the Forest Service by 1946. Clarence Forsling, assistant chief of the Forest Service before he was appointed to direct the Grazing Service in 1944, had hoped to rehabilitate Taylor lands by implementing range management principles like those advocated in the Forest Service, but he didn't get the chance; the service he directed was dissolved in 1946. Its management duties merged with those of the General Land Office in the new Bureau of Land Management (BLM), and Forsling lost his job. For decades afterward, the BLM remained committed to avoiding cuts in livestock numbers and passing along the costs of range rehabilitation where it was undertaken to the public at large rather than to the stockmen, who profited from relatively cheap range.[127]

The failure of the Forest Service to secure authority over more public rangelands in the 1930s is easily understood as a lost opportunity. As historian of Forest Service grazing policy William Rowley put it, "Clearly the opportunities of the Depression for an agency such as the Forest Service that specialized in the assessment, protection, and planned use of resources were not as great as might have been had the administration of all public grazing lands been extended into the Forest Service's experienced hands."[128]

Researchers had developed range management principles in the Forest Service but had not yet developed a discipline separate from Forest Service needs. The work of the Soil Conservation Service (SCS) during the years of Dust Bowl rehabilitation provided an entirely different arena for range management practice, relying much more heavily on extension work and private and public cooperation in the soil conservation districts established in 1938. But even the SCS became less popular when rain returned in the early 1940s.[129]

The work of the Grazing Service (which was regulatory, not theoretical or experimental) prevented further scientific disciplinary consolidation. Professional foresters had organized before 1900. Range scientists, on the other hand, did not form a society of their own until after World War II, at a moment when their "profession had no status or unity."[130] It took another few decades for range management to become a discipline in its own right and find its way into other land-managing agencies.

Range conditions on public lands have improved since then, according to several studies done in the 1960s and 1970s.[131] If range scientists—whether in the Forest Service or not—had been able to manage the forage crop as well as the animal crop on all public rangeland decades ago, would range conditions have been better?

The answer depends on how one calculates the difference in range condition as a result of the practices of stockmen and range scientists. Both had in mind the maximization of agricultural production; both understood the grasslands to be the source of the forage crop. Only range scientists advocated the maximization of uncut forage crops. From the range managers' point of view, stockmen who took no account of the reproduction of the forage crop were degrading an important natural resource and ultimately jeopardizing their own businesses. Had the range managers prevailed, the result would have been a maximization of the forage crop on public (and possibly private) lands. The quarrel between the two points of view was based on the question of which crop to favor (or to "manage for"), not on the question of whether there was a crop to cultivate in the first place.

Issues of range condition and the effects of grazing on private as well as public land are notoriously and perennially contested. Livestock interests have long been convenient villains as despoilers of the western range, and federally generated range management expertise is easily seen as a force mitigating the commercial abuse of grazing land. There is a good reason for this: livestock consume and trample the grass, while range managers attempt to restore and conserve it. Well-managed range and livestock appear to satisfy a demand for the adaptation of human activities to the "natural limits" of the environment.

More than any other landscape, grassland grazed by livestock lends itself to the illusion of "adaptation" because its identifying flora—the immortal grass—does not necessarily disappear with use, even commercial use. And in the event that native grasses are replaced with domesticated ones, a tame-grass pasture still looks more like the native range than like a field of row crops. In most lumbermen's calculation, the highest use of the forest means cutting it down, even if you replant it. You can't cut a tree a little bit this year and a little bit next year too. The highest use of the desert makes irrigation necessary, which entirely replaces desert vegetation where the water falls. The grassland is unique; properly cared for, with a well-adapted livestock industry, the grassland can sustain human use without ceasing to be a grassland even temporarily.

The perspective that views well-managed grazing on western grasslands as representative of sound environmental adaptation uses the condition of the grasslands (as the natural climax or its close approximation) to measure the degree of adaptation—that is, the degree to which recognition of environmental limits rather than people's notions of progress or improvement informs how people use grasslands. But if we consider the case of grazing animals, it is immediately clear that western stockraising did nothing to stop the progressive cultivation of nature. Moreover, in addition to cultivating their herds and flocks, stockgrowers eventually did get around to improving their pastures after the hard winters of the 1880s demonstrated the value of stored winter feed. At that time forage directly became a crop, and as in the forest a few decades earlier, all of the flora and fauna of the range came under the scrutiny of state scientists who attempted to transform the range itself into a crop-producing entity.

The story of range "adaptation" measures management progress against nature, even though this story is often told by those who otherwise dissent from a Turnerian or progressive view. For some writers who think of themselves as regionalists, nature remains fixed at the beginning of things, but instead of being the first point on a time line, nature is the basis on which all local activities depend and with which they must constantly negotiate. If nature is stingy in the matter of water, as it is in the subhumid West (at least in comparison with the humid East), then forms of land use will develop that are distinctive to that region's aridity. John Wesley Powell, Walter Prescott Webb, and Donald Worster are prominent examples of researchers holding this regionalist point of view.

For Worster, who uses Webb and Powell to define the West as a place rather than a process of settlement, the two modes of life unique to the West exist side by side rather than replacing each other in chronological sequence: the "hydraulic West," characterized by large irrigation projects, began as far as we know with the Hohokam and continues to be a major influence in western society and agriculture; the "pastoral West" did not precede the development of irrigated agriculture but was a relative latecomer after the introduction of cattle, sheep, and horses and has by no means disappeared. Western irrigation and pastoralism "must continue to obey nature's demands" concerning water, a condition that sets the West apart from other regions where the technologies of production are not so affected by environmental limits.[132] Citing J. Frank Dobie, Worster notes that the "history of any land begins with nature, and all histories must end with nature." That is, history must account for the effects of climate, vegetation, the presence or absence of water, the soil and topography, and the ecosystem or biosphere on any human activity.[133]

With respect to Plains historiography, beginning with nature means defining as Webb does the "physical basis of the Great Plains environment," those forces that are, "historically speaking, constant and eternal": geological formations, soils, climate, plant life, and animal life, in that order (followed as described in chapter 4 by the Plains Indians).[134] Instead of watching a parade of successively advanced settlers trail through the Cumberland Gap, we watch the progress of natural history as it lays the foundation of a specific region, the natural habitats of flora and fauna, which have their own histories.

Interestingly, for Webb the Plains Indians did not seem to represent a form of human adaptation to the Plains. They were found in the Plains environment—as particularly troublesome residents, even more so than Native Americans encountered in the East—but they were not native to the Plains in the way that the grasses and ground squirrels were, and they did not have an "approach" to the Plains as the Spanish and the Americans did. In Webb's account, Plains Indians were simply an obstacle to other people's "approaches." Plains Indians allegedly notwithstanding, successful human approaches to the Plains have used the region as pasture in some form, so much so that the "history of the Plains is the history of the grasslands." Webb wrote that the "physical basis of the cattle kingdom was grass," and the same was true for the fenced and well-managed stock farms he praised in 1931.[135]

This emphasis on the adaptation of human uses to natural western grasslands is central to the twentieth-century science of range management. Rather than a "primitive" stage in the history of agricultural progress, well-managed pastoralism becomes for a regionalist (or range manager) the highest, that is, the most ecologically stable and economically successful, form of adaptation to western grasslands—"a cultural as well as biological climax state," as Worster put it, that encompasses not only the ecology of grassland flora and fauna but also the human economy dependent on it.[136]

Both the progressive and the regionalist accounts of grassland pastoralism understand nature as the first element of history. A progressive point of view seeks to cultivate nature and transform it, as we know. A regionalist point of view seeks to "adapt" to nature, creating a dialogue between nature and society rather than a wholesale, one-way transformation. But the two apparently opposing perspectives remain representationally entangled in one another, as in Frederic Clements's use of the image of pioneer settlement to illustrate the theory of plant succession so fundamental to range management practice. Even without such an explicit reference, the idea that a primitive plant society gives way to successive stages of development, culminating in a climax formation understood as "adult," gives a thoroughly naturalized expression to an otherwise social phenomenon. It should not surprise us that an imperialist nostalgia would understand history as a "natural" process at the same time that theorists of nature would refer to social "progress" as analogous to natural history. In any case, the so-called adaptation of western stockraising to the environment took place in progressive agricultural terms, under the sign of improvement and the maximization of commercial production.

To some readers, the displacement of cultivation from the land to the bodies of cattle and sheep and the management of stock and pasture according to ecological principles spelled out by range scientists may represent a satisfactory (or potentially satisfactory) adaptation to the grassland environment. Who cares, we might ask, if people cling to their ideas about progress, as long as they don't actually plow up fragile marginal lands or allow their animals to eat the grass down to the dirt? What difference does it make?

The difference is that a regionalist perspective that fails to acknowledge the durability and near ubiquity of the progressive narrative may fail to address the continuing authoritarian enforcement of processes of domestication by people working toward otherwise "sustainable" stockraising. This domestication, we must remember, is as social as it is environmental.

The experience of Navajos at the hands of the Indian Service under the leadership of John Collier in the 1930s makes this point clear. Navajos were compelled to implement programs of stock reduction and breed improvement that focused exclusively on the material condition of the range, the uselessness of their horses (in terms of commercial value), and the degeneration of their flocks and herds (again, in terms of commercial value)—an "objective" problem of range condition and carrying capacity—rather than on the many social and economic factors involved in Navajo stockraising. Indeed, as Ruth Roessel wrote, "The hundreds of thousands of livestock killed or removed from the Navajo Reservation is but one, and not the most important, part of the drama."[137]

Stock reduction directly and profoundly threatened many Navajo families' livelihood. "At first, goats and sheep were taken away from the *Diné* [Navajos] and driven off somewhere. Soon they filled corrals at every trading post, and lots of animals just starved to death right there. The *Diné* told each other to eat some sheep and goats; so a lot of butchering took place. . . . So, John Collier caused our sheep to disappear, and then hunger struck us. Some *Diné* almost starved to death. He had taken a large part of our sheep, goats and horses. And he acted just like this was nothing."[138]

The reductions were enforced by federal marshals and the Navajo police, who arrested and jailed resisters. A Navajo interpreter presented papers to Navajo farmer Martin Johnson: " 'This paper says that you will let loose 75 head of goats as of today. If you don't obey the order, as it is stated, tomorrow the police will come again, and if you don't let go of the 75 head of goats, you and your wife will go to jail until you agree to do so.' "[139] At the Fort Defiance jail, the "cells were packed, and prisoners kept coming in. There was lots of commotion going on, and disturbance. The real reason why these people were jailed was simply because they did not want stock reduction and grazing regulation. . . . They fed us very small amounts of oatmeal with a few drops of milk. They would throw the bowls into the cell . . . just like feeding the dogs."[140]

Another man, Capiton Benally, complied with stock reduction for several years but finally had had enough. Benally's narrative indicates how tension rose on the reservation in direct response to the authoritarian process of stock reduction: "Then they asked me to reduce my livestock again," he said. "I didn't like it. Many Navajos were still afraid [of police persecution], and they went on the move away from the policemen." Benally had been told to inform another man that he, too, had to reduce his stock, but instead Benally told him, " 'Don't give your sheep away to them. Let's go to the supervisor and beat him up because we're suffering by being treated this way.' "[141]

One solution to the crisis was the proposed expansion of the reservation by 2 million acres in exchange for stock reduction, but this compromise was not realized, although many Navajos believed compliance with stock reduction would guarantee more land. "Today, these people are still saying, 'We were deceived, we never got the land we were promised.' "[142] Moreover, many Navajos rejected the idea that the deterioration of range conditions on the reservation was due to overstocking, which resulted in the "incomprehensiveness of the reduction as seen from the Navajo point of view."[143]

The Navajo experience with federal plans for range rehabilitation further illustrates the complex social aspect of these regional, environmental, and material problems. The first large-scale rangeland rehabilitation project undertaken by the Soil Erosion Service (SES) was directed not at any of the 330 million acres of private rangeland known to be in poor condition in the 1930s[144] but at the 2 million acres of depleted range on the Navajo reservation. Hugh Bennett, the first director of the SES (and the Soil Conservation Service that replaced it) had in fact been working with Indian commissioner John Collier on a report of Navajo range depletion before his appointment to the new SES in 1933.[145]

Navajos understood that the coercion employed by the Indian Service in the implementation of SES work was not just about stock reduction but involved the enforcement of authoritarian power at the expense of either local control or any sort of negotiated settlement. The SES staff working on soil stabilization, irrigation, reseeding, and rodent control, in addition to stock management, employed 85 college-educated Anglos, who commanded 700 Navajo workers. Given the opportunity, Navajos not only resisted Collier's and Bennett's stock reduction program but also rejected the Indian Reorganization Act altogether in 1935. The Navajo Tribal Council, which had approved Collier's stock reduction program only in the face of Collier's threat to implement it by force if the council failed to assent, ceased to function at that time as well.[146]

Not only were the federal programs implemented coercively on the Navajo reservation, but the programs themselves were informed by progressive ideas about the "improvement" of Navajo stock that favored commercially valuable qualities over all others. It was not enough that the Navajo reduce their stock; they must also privilege commercially valuable animals like sheep, cattle, and goats over socially valuable animals like horses (to federal bureaucrats, Navajo horses were simply "worthless"),[147] as well as improve the commercial quality of animals by breeding them systematically with purebred stock brought in by the Bureau of Indian Affairs, the Bureau of Animal Industry, and the SCS for that purpose.[148]

SCS range managers assumed they knew best how to manage the range and the people who made their living on it. Senator McCarran understood the implications of this authority well enough, at least as far as it affected the stock interests he represented. And the ideological model that informed range managers' practice was profoundly agricultural: it was intolerant of nonagricultural uses of the land, including the failure to improve a resource as well as the apparent "degeneration" of unmanaged land. A range manager "managed for" the range, not for the variety of social factors, including colonial racism and capitalism, that brought people, animals, and forage plants into often hierarchical relationships with each other on western rangelands.

Many regionalist studies of western land use and land and water laws, beginning with John Wesley Powell's *Report on the Arid Lands*, have emphasized the ways in which American consciousness, institutions, and agricultural practices had to adapt to the western environment, or how desirable such adaptation would be. But the progressive idea that nature must be cultivated, and that primitive methods must be replaced by scientific ones, was unaffected by western aridity.

The combination of grassland, herbivore, and human society as an illustration of social adaptation to nature has an attractive precedent in the relationship between grasslands, American bison, and indigenous Plains societies before the 1870s. Donald Worster cites several other examples, notably the Swiss community of Törbel, where people have been stocking local pastures for almost 800 years without depleting them, and without enriching some residents at the expense of others, under a system of community regulation.[149] Recognizing that western ranchers, federal scientists, and bureaucrats are still struggling to establish "the best fit between livestock and the land," Worster understands that it is the "fit" that is important, not pastoralism alone, and that many pastoral societies have destroyed the basis of their economy in the absence of a sustainable relationship between "livestock and the land."

But the sustainability of this relationship is never only a matter of animal and plant biology. The success of Törbel is social as well as environmental. We should not assume that pastoralisms fail merely because the grass gives out, but instead should ask why a range was overstocked, why a society's range territory was diminished, how people determine the "value" of their livestock. Just as the conservation of the forest involved a proliferation of colonial power over it rather than any limitation of the progressive impulse to control it, the "adaptation" of western agriculture to the environmental "limits" of aridity amounted to a proliferation of sites of improvement and control, including the management of animal biology and pasture composition on the part of stockgrowers and the condition of rangelands on the part of state scientists. The experience of Navajo sheepherders and rug weavers indicates how profoundly cultural imperatives informed the work of the "experts," at the Navajos' expense and at the risk of destroying pastoral expertise already endangered by the reservation system. What the experts had to offer was a limited and limiting understanding of plant and animal value, expressed in ever more complex systems of agricultural management.

Weeds

Monocultures first inhabit the mind, and are then transferred to the ground.
—Vandana Shiva, *Monocultures of the Mind: Perspectives on Biodiversity and Biotechnology*

Amateur botanist William Darlington wrote in 1847, "As the aborigines disappeared with the advance of the whites, so do the native plants generally yield their possession as cultivation extends, and the majority of the plants to be met with along the lanes and streets of villages, and upon farms, are naturalized strangers, who appear to be quite at home, and are with difficulty to be persuaded or driven away."[1] Darlington's account of native and introduced plants is much like Alfred Crosby's more recent description of the history of European plants in the Americas: in Mexico, "the cultivated fields of the declining Amerindian populations reverted to nature, a nature whose most aggressive plants were now exotic immigrants"; in California, the native "flora was as fatally vulnerable to European invaders as were California's aboriginal peoples, but isolation protected the flora, as it did the people, for two and a half centuries after the first coming of the Spanish to America."[2]

Crosby's account sounds more realistic than Darlington's—who, after all, persuades a weed to depart?—but both writers describe a basic intimacy between plants and people. More specifically, both writers describe the relationship between human immigration and plant immigration, between European human invaders and their accompanying plants, which were themselves aggressive immigrants. Perhaps Darlington's image of the naturalized stranger who cannot be "persuaded or driven away" best captures this relationship. The plants of the colonizers form such a significant part of what the colonizers are that they can be described as if they were human colonists and invaders themselves. Crosby counts European plants among what he calls the portmanteau biota of European explorers and colonists, his "collective name for the Europeans and all the organisms they brought with them."[3]

Crosby's perspective, as his book's title, *Ecological Imperialism: The Biological Expansion of Europe, 900–1900*, suggests, is biological. Regarding the human and floral invasion of the American West, the perspective I want to elaborate here is social. European people were organisms that formed part of an invading biota. Likewise, exotic plants were fundamentally part of the social phenomenon of invasion in the American West (and elsewhere).

Weeds become objectionable not because they are inherently ugly or useless, or because their growth is rapid and unchecked, but because they take territory and profit away from agriculture in some way. This may seem obvious—certainly every society must deal with weeds that interfere with food production—but the *form* of food production that developed in the West guaranteed a short list of useful plants and a growing list of weeds and determined how losses to weeds would be described and controlled. The specialization of western agricultural commodities relegated hundreds of plants to archaic and forgotten medicinal, ornamental, as well as culinary traditions. The elimination of even those plants that had long been acknowledged pests of crops became much more important because the midwestern and western farms they grew on were bigger and entirely dependent on the commercial value of crops they produced. And in fields whose only value lay in the profit made from the sale of crops, the presence of weeds could be expressed as a loss of productivity measured in dollar amounts and the control of weeds likewise could be measured as part of the cost of production.

More acres plowed and planted with imported wheat or alfalfa (and their companion weeds from across the Atlantic), more overgrazed and eroded rangeland, and the large volume of seed moving from farms to railheads and from railheads to destinations throughout the country together guaranteed weeds many opportunities to establish themselves and spread on an unprecedented scale. The response would take place on the same scale. State and federal legislation attempted to stop the widespread introduction and movement of weed plants and seeds; agricultural scientists developed programs of mechanized weed control necessary on large farms where centuries-old methods of hand-weeding were impractical; and other weed specialists began a chemical campaign against weeds that culminated in the development of 2,4-D in 1941. Before turning to the history of weeds in the American West, however, it is important to understand the significance of weeds to agriculture more generally.

When Darlington wrote his 1847 manual of weeds and useful plants, he intended it as a farmer's flora—as reflected in its original title—an ordered list of all the plants a farmer should know about, including those cultivated by accident. He saw himself as a botanist whose intellectual rigor enhanced the broader project of developing agriculture as a science. He was not writing to "the many plodding disciples of the old school of Agriculture, who despise every form of knowledge derivable from *Books*," but to the next generation of farmers who should be "taught the importance of scientific precision."[4] The exclusivity of the scientifically managed field is clear in Darlington's botanical philosophy: there are weeds and not-weeds, and it is the job of the farmer, armed with the best botanical information available, to favor the latter.

But, of course, the farmer has to "present the most favorable conditions possible" to weeds and not-weeds alike at the outset. Consequently, the "labors of the agriculturist are a constant struggle."[5] By understanding the manner of growth and mode of propagation of weeds as well as "useful" plants—that is, by becoming familiar with the biology and botanical classification of all of the plants he is likely to encounter—a farmer can stop the weeds from growing and "make certain plants grow and produce to their utmost capacity."[6] The preoccupation with that "utmost" would lead to standard calculations of losses due to weeds in twentieth-century weed identification and control reference works. And the constant struggle of the agriculturist became institutionalized as a "battle" against weeds, waged by any means at hand to exterminate or eradicate them and justified by the calculated value of lost agricultural productivity.

In 1911 Iowa experiment station botanist L. H. Pammel attributed a "crop shortage on many farms in this country" to weeds and suggested that farmers everywhere could increase their crop yields over 30 percent by preventing the growth of weeds. In Iowa alone, he wrote, weeds in cornfields cost farmers $7–9 million annually, estimating the total value of crops lost to weeds in the entire country at $100 million. Weeds are "injurious" to crops in many ways: they exhaust the soil nutrients and water needed by crops; they make it difficult to harvest crops; they harbor plant diseases; and they are sometimes directly injurious or poisonous to people or livestock. Pammel described all of the means of dealing with weeds under the chapter heading, "Extermination of Weeds,"[7] indicating the degree to which weeds should be excluded from not-weeds. Cornell botanist Ada Georgia reiterated many of the same views on losses and injuries caused by weeds in his 1914 *Manual of Weeds*. "Altogether," he wrote, "the losses sustained by the American farmer from this cause are greater than he suspects or would believe. A needless loss too," he continued, "for there is no weed so vicious that it cannot be subdued, with profit to the owner of the soil, if its habits are well understood and sufficient determination goes to the battle."[8]

Walter Conrad Muenscher was a Cornell botanist who wrote for a more scientifically educated audience than Georgia's (the latter's book was published in Liberty Hyde Bailey's Rural Manuals Series, whereas Muenscher's 1935 book was in Bailey's Rural Science Series). Muenscher consequently expressed many weed control concepts in less popular, more scientific language but nevertheless used Darlington's remarks about the agriculturist's "constant struggle" with weeds as an epigraph to his book. Muenscher used principles of plant ecology to describe this struggle in contemporary terms: "Weeds compete with crops for water, nutrients and light," he wrote, translating the farmer's struggle to encourage one set of plants at the expense of another into a competition among the plants themselves that justified the exclusion of the unwanted competitors.[9] He described the "three fundamental objectives of various methods of combating weeds" as prevention, eradication, and control. Prevention involves excluding weeds from "areas not yet infested, or preventing their spread from infested to clean fields," bringing the vocabulary of hygiene to the issue. Eradication "involves the complete destruction of a weed" once it has become established, and "control" interferes with the growth or reproduction of weeds in cases where eradication has failed. But eradication was still the goal: "Sometimes after control measures have been applied to a weed consistently for several years, it may be reduced in numbers, area or vigor so as to make its complete eradication practicable."[10]

In the 1940s, the weed problem was expressed in similar terms, only it had become more expensive. "Competition between crop plants and weeds is a most fundamental factor in the growing of useful plants," wrote a team of weed scientists in 1942, like Muenscher relying on the ecological principle of plant competition to explain the relationship between weeds and crops and the imperative to remove weeds. "A conservative estimate of 3 billion dollars has been given as the annual loss in the United States because of weeds," they wrote, without adding money spent on weed research, seed inspection, seed laboratories, weed law enforcement, and eradication programs.[11] In 1940 scientist and educator Edwin Spencer wrote, "In the struggle for existence a bad weed is a prince. It has the traits of a Bonaparte or a Hitler. Give it an inch and it will take a mile." For him, "the farmer, the truck grower, the gardener, the orchardist, and even the greenhouse keeper must wage a continual war with these persistent plants."[12]

Ada Georgia, no less a scientist because of his tremendous facility as a writer, expressed the philosophy of weed exclusion perhaps most comprehensively. He wrote, "Nature is the great farmer. Continually she sows and reaps." But the apparently indiscriminate sowing and reaping of Nature is always in excess of what "man" wants and needs, and while she "seems to yield a willing obedience, and under his guidance does immensely better work than when uncontrolled," it is "only by the sternest determination and the most unrelaxing vigilance" that her fellow worker can "subdue the earth to his will and fulfill the destiny foreshadowed in that primal blessing, so sadly disguised and misnamed, 'Cursed is the ground for thy sake; in sorrow shalt thou eat of it all the days of thy life; thorns also and thistles shall it bring forth to thee; and thou shalt eat of the herb of the field.'" But, he continues, "the civilization of the peoples of the earth is measured by the forward state of their agriculture; and agriculture in its simplest terms is the compelling of the soil to yield only such products as shall conduce to the welfare of the people who live upon it. It resolves itself into a contest with nature as to what plants shall be permitted to grow, and the discovery of the easiest, surest, and most economical means of securing a victory in the strife."[13]

Georgia underscores all of the themes touched upon in these chapters. Agriculture is a sign of civilization, undertaken by "man" as opposed to men and women, understood in terms of a domestication of a feminized Nature who must be compelled to yield only useful products. Georgia adds the element of thorns and thistles, however, that great menacing margin of agriculture where things are not only undomesticated but will remain so—never part of the "herb of the field" but never extinguished either, an object of vigilant and unending struggle. This agricultural philosophy rests on a primary distinction between thorns and herbs and defines agricultural progress as the exclusion of the former. This is true also for weed scientists more contemporary (and less literate) than Georgia who reiterate the drama of competition between crops and weeds rather than between man and Nature. In either case, agriculture divides the plant world into weeds and not-weeds and prescribes a program of exclusion for the weeds.

This distinction has nothing to do with the individual identities of the plants involved, as we shall see. It is a distinction that indicates the narrow instrumental aims of agriculture, which can only grow one thing at a time for a well-defined commercial market. A weed is anything that interferes with the uniform identity and value of a crop, regardless of its use elsewhere or any noncommercial value it may have.[14] The weed/not-weed divide is simple only with respect to determining the not-weed, a single plant identified and cultivated. With respect to identifying the weed, it becomes an untenable distinction.

The dictionary definition of a weed immediately reveals the complexity and ambiguity of the nature of weeds. According to the *Oxford English Dictionary*, a weed is a "herbaceous plant not valued for use or beauty, growing wild and rank, and regarded as cumbering the ground or hindering the growth of superior vegetation." Various forms of the word have been in use in English since at least the ninth century, but as the dictionary notes, its "ulterior etymology is unknown." The given definition is nevertheless sufficiently rich in naming the qualities systematically associated with weeds, however imprecise these qualities may be. First of all, a weed lacks usefulness or beauty, a judgment that can only be made in comparison with what plants an observer sees as useful or beautiful. Second, a weed is wild and rank rather than domesticated and controlled—that is, it does not flourish through the domesticating agency of a person but through its own agency, hence the disdain heaped on farmers or homeowners who allow weeds to grow on their property. A weedy field, garden, or lawn is a sign of neglect on the part of the owner in domesticating his property. The weeds, rather than the landowner, control the space where the weeds grow. Third, weeds take up space that should be occupied by plants considered superior in usefulness or beauty, and weeds may actually prevent the desired plants from flourishing.

All of these qualities depend on a value judgment that regards undomesticated plants as inferior to domesticated ones (a judgment we have already seen with respect to grass and cropland, feral cattle and "improved" cattle) and establishes the relative usefulness or beauty of a given plant, relegating those "without" use value to an inferior position in a hierarchy of plants. This is a progressive attitude toward weeds—that is, an attitude that values domestication over ferality and ranks plants according to their usefulness to people without acknowledging that what makes a cultivated plant "superior" originated in wild plants considered "inferior."

We have seen this logic at work in every project of "improvement" undertaken by agriculture. Weeds, like "wild" grasses, longhorn cattle, and "scrub" sheep, are excluded from the domesticated western landscape. The definition of a weed is further complicated by the fact that it depends on somebody's estimation of how useful or beautiful a plant is, whereas the same plant can easily be a drug, a crop, a salad green, a genetic resource, and a field pest to different people at various times or places. This ambiguity of "weediness" is widely acknowledged in literature about weeds, where writers often settle for defining a weed as a "plant out of place."

Moreover, as contemporary amateur botanist Sara Stein notes, a "weed is a plant that is not only in the wrong place, but intends to stay."[15] This vegetable intention is expressed as an "opportunistic genotype" in the language of weed specialists, who remind us that "weeds as a whole are less a botanical than a human psychological category," but weeds nevertheless share some biological, physical, and ecological properties.[16] Plants that can adapt rapidly and grow vigorously in several different environmental conditions and ensure their reproduction either by producing a large volume of seed or by the spread of creeping parts that give rise to new plants (or both) are likely to become weeds. These characteristics are generally shared by "pioneer" plants in ecological succession, which move into an area that has been disturbed by drought, flood, glaciers, fire, mudslides, or earthquakes. A plant with all of the possible physical characteristics of a weed "would probably take over the world" if people prepared its habitat well enough.[17]

It is, in fact, the role of people that complicates the biological life of weeds, just as it complicates the definition of a weed. Although many "pioneer" plants evolved in response to the physical disruptions of the Ice Age 2 million years ago, it has been human activity, specifically food production, that has provided the major habitat for weeds worldwide.

Furthermore, as people nurtured incipient crops, they selected and nurtured weeds as well. Stein explains the case of weeds in grains:

> The same selection to which farmers subject a domesticated crop also affects the weeds among them. Those weeds that happen to ripen when the grain ripens, and fail to shatter [that is, fail to drop their seeds when ripe, thus mimicking domesticated grains], are harvested with the crop, and threshed, winnowed, stored, replanted. Sieving the grain to remove weed seed selects for those that are the same size as the crop seed. Scrupulous weeding further intensifies selection for mimicry: those weeds most like the farmer's crop are those he is most likely to miss. In this way various wheat weeds evolved under the farmer's careful hand, including the mimicking tares whose bitter grains make bitter bread.[18]

Weed specialists refer to weeds' adaptability as "plasticity," which is "sufficient to allow for a rapid development of weediness when the chance for aggressive colonization arises."[19]

This anthropomorphic aggression, ubiquitous in the literature of weed identification and control, indicates much more than either a lapse of objectivity on the part of the scientists or the degree of undesirability of a weed. It indicates how closely people and weeds live together, as active parts of each other's activity. The way people talk and write about weeds (as well as insects, although we won't take up the history of western insects here) belies the lack of subjectivity and agency thought to be characteristic of scientific descriptions of the nonhuman world. Weeds are part of a portmanteau biota for Crosby; for Stein, they are commensals, eating at the same table with us, like rats and roaches, along with more welcome guests (presumably human and nonhuman alike). Ultimately, "weeds are too changeable to be snared by definition. They drift in and out of cultivation, enjoy popularity, drop out of favor, escape, hybridize, return as weeds, bend again to domestication. Wild sunflowers in old America became domesticated sunflowers under the nurturing of Indians, and weed sunflowers in the path of pioneers, and garden sunflowers at the hands of horticulturalists, and would turn purple if they had to in order to compel our care."[20]

The history of weed introductions around the world is no less complex or enmeshed in the history of human beings than the definition of weeds. "It has been said that the history of weeds is the history of man," wrote one team of weed researchers.[21] This is because weeds go where people go, whether by human design or not, leaving a floral record along paths of invasions and migrations. This record is legible but sometimes only with difficulty. One may be able to determine where a plant came from but not how it got elsewhere. Some of the immigrant plants in North America, for example, arrived with the knowledge and desire of the human invaders—clover and timothy, for example, which we have already discussed. Many plants that came from Europe were not crops at all, however, or even ornamental plants that had "escaped" from the garden. They were mixed in with crop seed; they were emptied onto shore in the ballast of ships or were dumped at sea and floated to shore; they rode in the wool of sheep, the hair of cattle, or the manure of animals shipped overseas, which was cleaned out of the vessels on reaching port; they traveled in the cheap straw used for packing crockery. Once here, these floral immigrants took up occupations related to those they had in Europe, as weeds of farms, gardens, dooryards, dumps, and other places disturbed by people, and spread along railroad beds and highway rights-of-way. Knowing where a plant comes from does not guarantee knowing its means of travel or its first destination.

The weed historian's difficulty in tracing the route of a plant's introduction is compounded by the difficulty of knowing which plants were introduced in the first place. "It is by no means easy to determine whether a plant owes its position in the flora of a country to Natural or Human agency," wrote Henry Ridley in his authoritative 1930 account of worldwide plant dispersal.[22] Botanists interested in weed histories are generally concerned with those species of plants introduced into colonized areas within the last 300 years or so—many of them, like the people who brought them, "pioneers"—which make up the bulk of undesirable plants.[23]

Histories of weeds and weed introductions rely on the records of explorers, soldiers in invading armies, settlers, and plant collectors to determine what plants were found in a given area at different historical moments. These histories rely on a tradition of botany begun in the sixteenth century in Europe, where plant collectors first began compiling ordered lists of all the plants they could find. Swedish botanist Carolus Linnaeus, born Carl von Linné, built upon this work in the eighteenth century when he devised an elaborate system of plant classification based on the physical anatomy of plants. Although a flora based on collectors' specimens and botanical analyses might appear to satisfy every demand for "objectivity," collectors only collect a plant when they see it in a particular place at a particular time—they often have no way of knowing how it got there or how long it has been there, only that it was there at the same time they were, and they have no way of knowing what plants have been overlooked. Also, the system of classification is always subject to change. A plant may have been given a Latin species and genus name in the eighteenth century only to be given a succession of other names up to the present. All of this botanical and historical confusion led Alfred Crosby to use only the common names of plants in his chapter on weeds in *Ecological Imperialism* "for fear of giving an air of exactitude to what must be, no matter how freely I resort to Latin and Greek, an imprecise account."[24] Common names vary widely as well.

Of course, organizing and naming plants according to their parts is not the only way to compile information about them, although it is the primary scientific system of classification used by botanists. (The primary social system of plant classification, no less important to weed scientists and other people, is based on the working distinction between a weed and a desired plant.) Since no flora as such were made in places like North America before European explorers, plant collectors, and colonists arrived and therefore many introduced plants had decades or even centuries to spread before being collected as herbarium specimens, weed historians cannot always know which plants are native and which were introduced. Moreover, those plants classified as "native" have been no less involved with human work along the routes of wandering and conquering peoples for the last 10,000 years at least, wherever people carried means of food production with them.[25]

Although a plant may be indigenous, it is nonetheless likely to have been part of some society's history and movements for a long time. A plant may be native (and even uncultivated) without being part of a "nature" separate from society, accounted for in the economies and knowledge systems of people whom botanists often neglect to consult. "The historical record is simply not clear" with respect to weeds, C. L. Foy, D. R. Forney, and W. E. Cooley admit, but it is nevertheless compelling testimony to the intimacy of plants and people, including weeds and colonizers.[26]

With this general understanding of weeds and their history in mind, we can look more closely at their tenure in the West. During the mid-nineteenth century, weeds continued to spread, as they had done since European occupation in the Southwest and on the Atlantic coast. Some of them had already become widely naturalized across the continent. One of these was broadleaved plantain (*Plantago major*), whose history demonstrates both the relationship between colonization and plant introduction and the propensity of a weed to get away, in this case, to get away from an official, colonial history to become part of a history of other societies.

A curious story about plantain appears frequently: it is said to have followed the white man so closely in his movements that "the Indians" called the plant white man's foot or Englishman's foot.[27] Of course, the Indians in question are never identified more specifically. The scientific name is not derived from any of the North American languages but from Latin, referring to the sole of the foot in the language of the Romans. A less common conjecture regarding plantain is that the Romans named the plant after the sole of the foot because it appeared to follow *them*, growing wherever "the Roman legions . . . trudged to conquer the known world."[28] Darlington implied the same history when he wrote that plantain was "remarkable for accompanying civilized man—growing along his footpaths, and flourishing around his settlements [hence another of its common names, dooryard plantain]. It is said our Aborigines call it 'the white man's foot,' from this circumstance."[29]

Whether Native Americans across the continent recognized the relationship between plantains and white exploration and occupation and then named the plant white man's foot or whether settlers identified themselves with the legendary conquering and civilizing force of the Romans by renaming the plant Englishman's foot is uncertain. Plantain in any case became one of the earliest floral signs of European contact and successful conquest in the eyes of Euro-American writers and botanists, although presumably it took up residence in any hospitable dooryard or footpath. Like the European immigrants with whom it was originally associated, plantain is so widespread and its continental career has been established so long that some writers deemphasize or deny its exogeneity. "Broadleaf plantain is supposedly a native of Europe; but today it is cosmopolitan in distribution," wrote a group of weed specialists documenting western weeds.[30] The USDA Agricultural Research Service notes equivocally that plantain is "probably [a] native of North America" as well as Eurasia, and herbarium curator H. D. Harrington included plantain in his list of edible plants native to the Rocky Mountain region.[31]

Plantains established themselves so widely and had been around so long that many different Native American societies, in the East as well as the West, developed their own uses and names for the plant. One of these was snakeweed—*omikikibug* in the Algonkian language of the Anishinabe; *ginebiwuck* in the Siouan language of the Omaha—because poultices made from the leaves were used to treat snakebites.[32] The Cherokee and Mohegan also used the plant to treat snakebites.[33] At least eighteen societies—ten of which lived from the northern Great Plains westward—have used broadleaved plantain to treat a variety of conditions, including wounds, fevers, earaches, sore eyes, diarrhea, and colds. Among the Navajo, the leaves have been used to treat a condition they have identified as lightning infection as well as for ceremonial purposes.[34]

Wide use of the plantain in North America as a wound dressing and local adaptations of that use (such as, for example, treating lightning infection) suggest a set of associations with the plant entirely different from the associations held by Europeans, for whom plantain had become by the nineteenth century merely a dooryard weed with apocryphal reference to colonists and the steady march of civilization. The textual identification of plantain as the foot of conquerors and the apparent geographical ubiquity of this useful, life-sustaining plant reveal a profound limitation of the Euro-American understanding of the plantain and its history. The language of Anglo plant historians presumes a definite conquest by identifying the range of plantain with the range of the conquerors' power, an identification belied by the fact that Native American people have used the plant for their own purposes apparently wherever they found it. It was not a "weed," although it was not necessarily cultivated, and non-European plant gatherers were free to take it as their own in such a way that it no longer belonged to Europeans or a European medicinal tradition. Plantain was also certainly not a harbinger of cultural "defeat" but became a plant with sustaining material and spiritual uses entirely separate from the geographical and cultural context it came from.

Plantain provides a good example of a plant whose history confuses the natural or "native" with the exogenous or "cultural" in many ways. Once a weed becomes "naturalized," like a human immigrant, presumably it no longer matters where it came from; it only matters that it is here now and flourishes as if it had always been here. But naturalization also means that weeds, like people identified with nature, become part of an undomesticated wilderness, a force of nature that must give way before the tools and civilization of a culture. The naturalization of a European plant erases its traditional uses—as a medicinal herb, a salad green, a forage crop—as well as the practical memory of the plant as useful or socially significant. In the case of plantain, Euro-Americans largely abandoned using its leaves to treat wounds and stings; very few people can identify, let alone have any practical memory of, this plant that is still likely to share our paths and trampled yards. Once part of nature, a plant is no longer part of culture and is subject to either domestication or eradication.

Unfortunately for those who are staunchly allied with culture, many plants' histories and habits belie the exclusivity of the nature/culture divide, and these plants are incapable of simply succumbing to control or dying. Their vitality—always intertwined with human vitality—becomes their aggression, and their persistence outside of domestication becomes a pernicious or insidious threat to a stable agricultural order, when in fact it is agriculture that often introduces and spreads weeds.

In the case of human immigrants, naturalization involves a similar morass of social and biological categories and indigenous and exogenous categories. A citizen of one country is allegedly absorbed into the "natural" body of another adopted country (in the United States, of course, a "nation of immigrants" is imposed on the nations of indigenous peoples), erasing his or her former national identity. But the process of naturalization is a social one, whereby the person who is "naturalized" is actually supposed to be culturalized anew. The "nature" in which he or she becomes naturalized is an elaborate set of social expectations; naturalization attempts to erase one set of social allegiances, identities, memories, and functions in favor of another. The process of naturalization is complicated by perhaps unanticipated social values and expectations, for plants no less than for people.[35]

Imported through colonial contact and associated with conquest in the minds of the presumptive conquerors, plantain nevertheless behaves like a weed: it escapes its "official" history to become something quite different in other people's hands, a plant whose history remains linked with human history without being contained by any one manifestation of it and without being "naturalized." As we look at other plants introduced in the West and subsequent means of spreading and controlling them, we should keep in mind this larger, positive understanding of weediness: a propensity to persist, to escape exclusion, and to defy both "naturalization" and eradication in the face of conquest.

Weeds—most of them companions to colonization, if not celebrated for it—found many routes into the West as railroads stretched to the coast, as new territories recruited large numbers of immigrant farmers to land cleared of Native American inhabitants, and as the USDA began recruiting new crops from Europe and Asia for use on the Great Plains and the western range. As always, the movement and cultivation of crop seed "contaminated" with weed seeds was a common mode of unintended introduction of western weeds, from the relatively benign plantain to the more dreaded Canada thistle (*Cirsium arvense*) and leafy spurge (*Euphorbia esula*).[36] Plantain does not rank very high on the scale of perniciousness among western weeds. Darlington wrote, "It is rather a worthless weed, but not much inclined to spread."[37] Other introduced plants have become notorious for the inroads they have made on western rangeland and farmland and have pressed agricultural scientists, botanists, and weed specialists toward a vehemence not otherwise found in their literature. Canada thistle and leafy spurge are among the most loathed western weeds, and yet their histories, like that of the plantain, illustrate the complex and inseparable relationship between plants and their people.

According to the weed specialists who compiled *Weeds of the West*, Canada thistle arrived as a European contaminant of crop seed in eastern Canada in the eighteenth century.[38] An earlier source wrote that Canada thistle "is reported to have been found about the residences of French missionaries early in the seventeenth century. There is a tradition that it was purposely introduced into Canada by the French for feeding swine; but there appears to be no ground for this tradition. It is said to have been introduced into eastern New York with the hay and camp equipage of Burgoyne's army in 1777."[39] In any case, from eastern Canada, the plant spread west and south.

Before it was associated with Canada, Canada thistle was associated with cultivated fields more generally, a connection reflected in its species name, *arvense*. The genus *Cirsium* was once used to treat varicose veins, but like plantain, its medicinal value was lost to Europeans in North America. Medicinal use of this thistle had probably long been in decline in Europe as well, however. By the time it was named, *Cirsium arvense* was a weed of fields rather than a plant of an apothecary's office, and Linnaeus wrote that it was "the greatest pest of our fields."[40] What makes this plant so difficult to control in fields is the fact that it is a perennial with creeping rootstocks. Cutting it down does not kill it or prevent it from reproducing, even if it never produces seed; indeed, in some places, Canada thistle does not go to seed even if left untouched.[41] Pulling it up or plowing it fragments the roots, which, instead of injuring the plant, multiplies it rapidly because each fragment is capable of generating a whole new thistle. A mid-nineteenth-century English writer explained that, because of its "nature and pernicious effects," *Cirsium arvense* was called "cursed thistle," and Darlington wrote that it was "perhaps, the most execrable weed that has yet invaded the farms of our country."[42]

Canada thistle was apparently slow to reach the West. It is primarily a plant of fields rather than range, so the 1937 *Range Plant Handbook* does not even mention it. Had it followed cultivation west with farmers moving to Washington and Oregon in the 1840s, it probably would have appeared in plant collections made in the Northwest before the 1880s or in ethnobotanical information gathered about western Native American societies before the reservation period. But weed migration scholars studying introductions in Washington, Oregon, Idaho, Montana, and Wyoming found no record of it in those states before 1890.[43] One source notes that Canada thistle was found in California in 1879.[44] Canada thistle was most likely introduced, then, with the influx of settlers in the 1870s and 1880s that produced the cattle and sheep booms, many of whom eventually turned to raising grain or forage crops and would have provided a market for crop seed of which Canada thistle seed was a common companion.[45] No Native American groups west of the Anishinabe appear to have adopted it,[46] which perhaps could be explained by this plant's arrival during a period of social and ecological upheaval for western Native American societies who were increasingly being moved onto reservation lands and compelled to follow Euro-American agricultural and medicinal traditions. The transport of hay and movement of cattle and sheep in the West after 1870 may have contributed to the spread of Canada thistle as well, both factors apparently having played a part in its introduction to New York State from Ontario.

Wherever it was introduced, Canada thistle was considered "one of the most feared weeds in the United States,"[47] "infamous,"[48] an "aggressive weed,"[49] "perhaps the worst weed of the entire United States," without "a single virtue so far as man is concerned."[50] Of course, even this rogue plant found uses among other people, including the Mohegans and Montagnais (who have used it to treat tuberculosis) and the Anishinabe (who have used it as a bowel tonic).[51] In 1942 weed specialists wrote that throughout its "extensive territory," Canada thistle "maintains a reputation as an aggressive and pernicious weed."[52] Even after decades of herbicide development, Canada thistle remains "hard-to-kill," a "troublesome," "serious competitor" for cropland.[53]

Another western weed infamous for its pernicious, aggressive behavior is leafy spurge. Like Canada thistle, leafy spurge is a perennial plant that spreads by creeping rootstocks. It also reproduces by seeds, which are expelled explosively from the seedpods and then carried by animals, birds, insects, and watercourses, as well as by people in crop seed and hay. Unlike Canada thistle, leafy spurge was not known and feared as a weed upon its arrival because it had not been a field pest in Europe. When it was imported in ballast and as a crop seed contaminant, however, its natural insect enemies were not imported with it. At least one weed specialist identified *Euphorbia esula* as faitour's grass, although the *Oxford English Dictionary* does not specify which spurge was known by that common name. A faitour is someone who fakes illness, and the acrid milk of spurge (possibly leafy spurge) was used for this purpose, giving the plant an outlaw association early on.[54] Leafy spurge apparently came first in ballast to Newbury, Massachusetts, where it was found in a dump in 1827, and was spreading freely by the 1840s, already "likely to become troublesome" because of its running rootstocks.[55]

The introduction of leafy spurge in the West was probably due to the emigration of a large group of people from Russia to Manitoba, North and South Dakota, Minnesota, Nebraska, and Kansas. In 1874, 18,000 Russian Mennonites brought wheat seeds among their belongings, and no doubt Russian weed seeds as well, including leafy spurge, which was native to the region they came from. Midwestern immigrants continued to import seed grain, including large shipments from the Crimea, after 1900. In the 1930s, leafy spurge was not a "common weed in this country as yet."[56] At that time, there were two distinct centers of leafy spurge introduction—one on the east coast, the other in the north-central states, and it was scattered as well in Montana, Idaho, and Washington—but one botanist already warned that it "should not be permitted to become [widespread], for its creeping, horizontal rootstocks make it difficult to dislodge when once established" and its spreading patches "smother all weaker growths in its way."[57] The persistence and vitality of spurge had begun to take on a kind of anthropomorphic villainy, as its survival even in adverse conditions was seen as pernicious, underscored by the gravity of the agricultural problem caused by the plant and the violence ("smothering") it inflicted on crops.

By 1942, leafy spurge was "a serious agricultural pest" in Iowa, Minnesota, and the Dakotas as well as in Canada. Weed specialists continued to warn against its spread, noting that the dense root system of leafy spurge "enables it to choke out such crops as sweet-clover and alfalfa," extending deep enough to be "assured of an adequate and constant moisture supply" even in drought.[58] Weed specialists knew in the 1930s that leafy spurge could be partially controlled by allowing sheep to graze it or cutting it repeatedly, but this method of control was seldom adopted, even when the available chemical means of control were relatively primitive.[59] The recommendation to use sheep to control leafy spurge was reiterated through the 1980s, but "the use of sheep is not an accepted and widely used practice," according to a contemporary leafy spurge specialist.[60] After the development of modern herbicides in the 1940s, at least one western weed researcher in the 1950s simply hoped a suitable chemical could be found to kill leafy spurge instead of attempting to control the plant by other means.[61] In the West, leafy spurge is still "a serious weed because of its spreading nature and persistence" and remains "aggressive" and "extremely difficult to control."[62]

Many western weeds were alien both in the sense that they were exogenous to North America and in the sense that they were unknown to the mostly western European immigrants responsible for the first wave of weed introductions. Some doubly exotic plants went directly to the West, either with immigrants from central and eastern Europe (like leafy spurge) or with an increasing number of crop plants imported from the same region for sale to western farmers generally. These included several species of *Centaurea* and Russian thistle (*Salsola kali*).

Spotted knapweed (*C. maculosa*), diffuse knapweed (*C. diffusa*), Russian knapweed (*C. repens*), and yellow starthistle (*C. solstitialis*) all arrived near the end of the nineteenth century in crop seed, alfalfa seed from western Asia apparently being a common source for each.[63] The earliest collection of diffuse knapweed in the West was in an alfalfa field in Bingen, Washington, in 1907.[64] Spotted knapweed was collected first in Montana about the same time, and Russian knapweed (probably introduced in 1898) was collected in Washington and Wyoming between 1911 and 1920.[65] Russian knapweed was also called Turkestan thistle, reinforcing the connection between these plants and the origin they shared with new varieties of alfalfa.[66] Yellow starthistle was a known companion of alfalfa seed and common in fields newly planted with alfalfa. It was also the first of the *Centaureas* discussed here to be collected in the Northwest, having been found in a county in Washington before 1900.[67]

Although these *Centaureas* began their tenure in the West as weeds of alfalfa, they eventually branched out (as it were). The knapweeds were able to establish themselves on any disturbed soil, not just cultivated fields, and were therefore poised for introduction onto overgrazed or eroded rangeland along with downy brome (discussed below). Russian knapweed, the only perennial among these *Centaureas*, had the advantage of spreading by root as well as by seed, but all *Centaureas* are believed to inhibit the growth of other species found near them, essentially clearing the ground for more of their own kind.[68] Neither diffuse nor spotted knapweed is grazed much, Russian knapweed repels all livestock because of its bitter taste like quinine, and the "vicious spines" of yellow starthistle are actually injurious to livestock.[69]

None of these *Centaureas* found a place in the *Range Plant Handbook*, but other evidence suggests that they were becoming important range and roadside weeds in the West by the end of the 1930s. Their territory, although not as extensive as downy brome, was increasing steadily.[70] Russian thistle had become permanently established and abundant in some places, even if not found everywhere in its range, from North Dakota to Missouri and west to the coast as early as 1922.[71] Muenscher noted in 1935 that spotted knapweed was found in the north-central states and the Pacific Northwest and was "spreading in recent years."[72] The spread of these species combined with their inedibility and their tendency to kill or crowd out other vegetation resulted in a recipe for plants beyond control once they got established on rangelands, roadsides, and railroad beds. They might be kept out of fields by plowing them under repeatedly for a year or two,[73] but this was not possible on rangeland or rights-of-way. The advent of modern herbicides didn't change this situation. Widespread chemical application on rangeland still involves "prohibitive cost" because the chemicals used to kill these plants in smaller stands are too expensive to use in large quantities over wide areas.[74]

Knapweeds arrived with agricultural settlers and traveled the same routes they traveled, along roads and railroads—uncultivated but often disturbed thoroughfares—especially as they were being built. Many weeds followed railroads as fast as they were laid down in the West. Russian thistle, the archetypal tumbleweed, was one of these.[75] Brought to South Dakota in 1873 in Russian settlers' flax seed, Russian thistle was recognized by the USDA as a problem on the northern Great Plains by 1881.[76] The spiny leaves of full-grown plants could pierce the leather gloves of farm and ranch hands and cut the legs of horses in pastures where it grew. When mature, it broke free of the soil and shook out its many seeds as it tumbled overland, generating thistles wherever it blew in the dry plains environment where it thrived. Although it has been named several times by botanists (and, of course, many times by other people), the genus name by which it is known today—*Salsola*—is a Persian word for carpet, which the plants' innumerable seedlings resemble in the spring.[77] Russian thistle filled chinks in sidewalks in plains towns and occupied yards and vacant lots. It spread rapidly over land cleared of prairie grasses. Some people thought fences would stop the flight of Russian thistles, but the tumbling plants piled up along fencerows and blew over them, and the wind continued to carry the seeds of those plants stopped by the barriers.[78]

By the 1890s, Russian thistle had followed railroads and irrigation ditches all over the West as thistle seeds in grain were loaded and transported by train or fed to draft horses used in excavating irrigation canals. In the intermountain region, the overgrazing that allowed other weeds an entrance invited Russian thistle as well. Its ability to survive without much moisture allowed Russian thistle to become emergency forage on ranges in drought-stricken areas where nothing else was green in the 1930s. But as the *Range Plant Handbook* pointed out, it "cannot be considered a desirable forage plant on mountain ranges because livestock will not eat it if other and better forage is available."[79] Still, that it was palatable at all sets it apart somewhat from the knapweeds and Canada thistle; it was out of control (although post-1940 herbicides had more luck with this plant than with some others),[80] but it leaned toward a margin of weedy ambiguity where it becomes less clear which plants are weeds and which are not.

Although cases can be made on behalf of the usefulness of Canada thistle, leafy spurge, *Centaureas*, and Russian thistle, these plants represent an outlaw fringe acknowledged as useful or helpful in emergencies at best, if at all. Other plants have led much more ambiguous lives in western agriculture. Some of these demimondaines were introduced on purpose as crops, forage, or ornamental plants and subsequently escaped the fields, gardens, and ranges where they were kept. Others were introduced unintentionally as weeds but had been cultivated at some point in the past. The same plant could be simultaneously sown on purpose in one place and cursed as a weed in another. And some plants introduced on purpose have yet to be recognized as weeds, but their virtues are equivocal, and the language used to describe their value can easily be translated into the negative complex of pernicious vitality reserved for proven enemies of agriculture.

Several grasses occupy this shadow world of dubious weediness, including downy brome (*Bromus tectorum*), johnsongrass (*Sorghum halepense*), couchgrass (*Agropyron repens*),[81] and crabgrass (*Digitaria sanguinalis*), all of which were useful at one time or another in the West as crops, as forage, or in stabilizing eroded land. Crested wheatgrass (*Agropyron cristatum*) is a common western forage plant whose persistence and drought resistance may one day emerge as weedy properties. Kochia (*Kochia scoparia*) and yellow toadflax (*Linaria vulgaris*) are examples of ornamental plants that escaped the domesticated and domesticating garden to become weeds of fields or rangelands as well as what weed specialists call "waste places."

Downy brome is one of those annual plants that evolved in grainfields and pastures by mimicking more highly valued grasses without giving its seeds (which are armed with long awns or spikes) either to the granary or to livestock. Its common names as a result include cheat and chess, names given to any weed grass that has pretended to be a valuable grain. The strike against downy brome as a valuable forage plant is primarily the fact that it matures and becomes unpalatable to livestock very early in the season. Its awns can injure animals' mouths as well. Its genus name, *Bromus*, comes from the Greek word for oats, which bromegrasses resemble, and its species name, *tectorum*, pertains to roofs, since this grass was at one time common in roof thatch—plausible for a weed grass long associated with grainfields and their straw.[82]

One certain means of downy brome's entry into the United States was in cheap straw used to pack crockery.[83] But downy brome might also have arrived in crop seed, hay, straw bedding, animal hair, or any combination of these. Downy brome was introduced in the East but was "not greatly troublesome in Iowa or eastward" in 1911.[84] By that time, though, it had become a serious pest wherever large amounts of wheat were grown, beginning on the prairies and the plains during the last quarter of the nineteenth century.[85] The completion of the railroads in the 1880s encouraged agricultural settlement in the intermountain region, where homesteaders often planted wheat. Several weedy species of brome were collected in the Northwest beginning in 1882, and downy brome itself was established there by 1889, almost certainly having arrived in wheat seed.

It was in the intermountain region that downy brome escaped the environment of the grainfield in which it had evolved. Unlike the grasslands farther east, intermountain grassland was dominated by non-sod-forming bunchgrasses that were perennial but easily plowed and easily damaged by grazing. Intermountain rangeland was so badly overgrazed at the end of the livestock boom that the native grasses had largely been destroyed and the soil had generally been disturbed, allowing downy brome to begin taking over rangeland. By 1935, downy brome was a weed of rangelands as well as fields and had become "very common in the western states."[86] Farmland abandoned during the depression following World War I, as well as during the Great Depression, was rapidly occupied by downy brome.[87]

The condition that led to the escape of downy brome from western grainfields also led to the deliberate introduction of downy brome in an effort to restore fertility to overgrazed, exhausted range. The USDA planted downy brome at Pullman, Washington, in 1897 as a forage grass, an introduction from which it may easily have spread to other areas. Although it was sown on purpose in some places to reclaim western rangeland, it also served the same purpose accidentally in other places; if downy brome had not been available to occupy abandoned farmland in the 1930s, the effects of erosion in that period would have been markedly worse than they were.[88] Also, since it matured early, downy brome was frequently the most abundant forage available at the beginning of the grazing season on the range, and at least one seed peddler apparently sold it as a good forage grass in Oregon in 1915. Although it is not clear when downy brome reached Arizona and New Mexico, at some time it became a material used in ceremonial medicine by the Navajo, who probably also began to depend on it for early spring forage.[89]

This grass is now the dominant species in many areas, so far beyond the possibility of control, let alone eradication, that some range scientists recommend a change from managing grazing to encourage perennial grasses (the usual goal of range grazing management) to simply managing how and when stock graze downy brome.[90] The value of this grass (however limited), combined with its abundance on the western range, finally compelled livestock owners and range managers to give up an unequivocally negative assessment of its qualities.

Johnsongrass, couchgrass, and crabgrass are all similarly part-time weeds. The two perennials—johnsongrass and couchgrass—are both vigorous, aggressive, pernicious, and extremely difficult or almost impossible to control, spreading by creeping rhizomes.[91] Although specifically sown as forage, johnsongrass is poisonous to livestock if it has been affected by drought or frost. The root tips of couchgrass are sharp enough to actually pierce the roots and tubers of other plants, and it is believed to poison plants growing near it. A cotton-growing expert returning from Turkey in the 1830s brought several plants back with him to South Carolina including johnsongrass, used widely as forage in India, which became an important hay crop in the Southeast. It spread to irrigated fields primarily in the Southwest.

Johnsongrass has been generally recognized as both valuable and potentially troublesome. As one commentator put it, johnsongrass "is said to be a very fine grass for hay, and for this reason many southern farmers sowed it when it was first introduced. One man says that if a farmer has no intention of raising anything but hay on his place, and at the same time has no regard for his neighbors' rights, he might be advised to sow Johnson grass, but otherwise he should never so much as think of doing it."[92] It was nevertheless distributed by the California experiment station as a forage grass as late as 1884 and 1885.[93]

Couchgrass was not introduced on purpose (it arrived before 1751, probably in mixed grass seeds),[94] but like johnsongrass, it had a reputation as good forage at the same time that its potential to escape from cultivation was well established. "If it were put to a vote," wrote botanist Ada Georgia, "perhaps most farmers would name quackgrass [couchgrass] as the most obnoxious of its tribe; yet it makes good hay and two crops a year of it, is sweet pasture grazing which cattle eat greedily, its matted 'couch' of interlacing rootstocks make it an unsurpassed soil-binder in steep gullies or on road embankments where the ground must be guarded against 'washouts.' But," Georgia continued, "it is its very tenacity of life that makes it such a pest when it gets into cultivated ground. If it could be kept in its place, or were it not so hard to kill when it gets out of bounds, it would be a welcome friend."[95] A clearer statement of the imperative to domesticate or eradicate weeds could not be found.

Crabgrass was introduced by the U.S. Patent Office in 1849 also as a forage crop but was not planted widely. It was, however, one of the oldest cultivated grasses in Europe, and its seeds cooked in milk were a common food in Germany and Poland. Its name may be derived from "crop-grass" as well as a reference to the crablike appearance of its spreading stems.[96] When central and eastern European immigrants around the turn of the century began settling in the West, they brought crabgrass with them as a grain crop that was sure to do well even if planted late or on poor ground. They eventually gave crabgrass up in favor of corn and wheat, and their first crop remained behind as a weed. Although an annual, crabgrass developed a strategy of vegetative reproduction common to the most "aggressive" perennial weeds: it can generate roots along its prostrate stems; fragments of roots left in the ground can generate new stems; and stems torn out and left for dead on the ground can root and begin a new life. Its scientific name, *Digitaria sanguinalis*, refers to the reddish color of the fingerlike branches of its flowering panicle (hence it is sometimes called fingergrass), but farmer and biologist Edwin Spencer, writing for a popular audience, added, "If sanguine means ardent and filled with enthusiasm, that name fits this plant perfectly," and he called it a "weed of nine lives."[97] Again, the vitality that made this plant a desirable crop for those who ate its seeds condemned it to weediness among those who found it useless.

Just as a plant might be introduced on purpose only to be eventually considered a weed, some plants introduced in the West have properties like those understood as "weedy" but for now retain their more privileged status. One of these is crested wheatgrass, a relative of couchgrass, both of which were named for their resemblance to wheat. Crested wheatgrass is a perennial bunchgrass (it has no rhizomes and does not form sod) native to the steppes of Russia and southwestern Siberia. It was imported by the USDA beginning in 1898 as a grass well suited to the West,[98] where the usual perennial grasses introduced for hay and forage (timothy, bluegrass, and redtop) had not been as productive as hoped.[99] It was tested at an experiment station in South Dakota in 1906. By 1915 it had been sown successfully throughout the northern Great Plains.[100]

Crested wheatgrass is both drought-resistant and hardy in winter. It became a common forage grass in the Dakotas, Montana, Wyoming, Idaho, and eastern Washington and Oregon and was introduced farther south as well, usually at altitudes above 5,000 feet, in California, Nevada, Utah, Colorado, New Mexico, and Texas.[101] Its "abundant crops of seeds" make "good stands . . . easy to obtain," and it can be grazed about two weeks (or even a month) earlier than native prairie grasses in the spring[102]—all qualities reminiscent of downy brome.

Like downy brome, crested wheatgrass has to be grazed early in the spring, after which the livestock must be removed until new growth begins with cool weather and fall rains. Its early start makes crested wheatgrass unpalatable to livestock relatively early: "As the grass matures, it becomes harsh, and the protein content decreases rapidly," according to the same forage expert who highly recommended crested wheatgrass for western pastures.[103] Cold and drought do not kill it. It grows rapidly and sends its roots as deep as ten feet into the soil, where they have ready access to water even in the driest periods. Crested wheatgrass can "crowd out weeds where the moisture supply is limited" and presumably other plants as well.[104] Two of the weeds it was recommended to compete with were leafy spurge and yellow toadflax, both creeping perennials feared and loathed for their ability to crowd out other plants,[105] indicating that crested wheatgrass could perhaps as easily smother plants susceptible to spurge or toadflax. Like downy brome, crested wheatgrass "has proved particularly valuable for regrassing abandoned cropland," but unlike its less-favored cousin, crested wheatgrass "has become the leading grass for use in the Northern Plains for pasture, hay and erosion control."[106] The only difference, of course, is that downy brome did most of its soil stabilization and forage production outside the control and planning of conservation experts. But like other weeds generally, crested wheatgrass itself has come to be "a common escape from cultivation."[107]

Only the continued reiteration of the values of crested wheatgrass—always known to be limited—and the perception that this grass is not "out of control" prevent it from becoming a weed in the West. Many other plants with as much (or as little) use value that have escaped cultivation through their drought-resistance, vigorous seed production, and ability to crowd out other perennial plants have been unequivocally judged weeds. We might ask how forage experts would deal with the range and abundance of crested wheatgrass—which is at least as extensive as that of downy brome—if this plant should someday fall from grace.

Occasionally, plants introduced for their aesthetic or medicinal qualities get away from the garden to begin careers as escapes. Such may have been the case with Canada thistle centuries ago in Europe. It was certainly the case with yellow toadflax and kochia in North America. Toadflax was used widely in Europe as a dye plant,[108] was boiled in milk to attract and kill flies, and served as an astringent, purgative, diuretic, and skin ointment.[109] One of its common names, butter-and-eggs, refers to the yellow colors of its snapdragonlike flowers, which have been reason enough to cultivate the plant for some people. Yellow toadflax was introduced to Philadelphia from Wales by a Mr. Ransted (whose name became another of the common names of the plant) before the eighteenth century and was adopted by immigrants from Germany happy to find a familiar dye plant already growing in the North American colonies.

Toadflax adapted so well to its new environment that it eventually no longer depended on cultivation in order to spread, and by the middle of the nineteenth century, it had become a "vile nuisance" in pastures and meadows.[110] It is a perennial plant "very difficult to suppress" because of its creeping rootstocks. Moreover, it is not at all a forage plant: "Cattle dislike its taste and odor, and in pastures it is left to reproduce itself unmolested."[111] Like Canada thistle, yellow toadflax apparently took centuries to reach the West. No Native American western societies except the midwestern Anishinabe adopted it as a medicinal plant (the Anishinabe have used it as a bronchial inhalant in sweat lodges).[112] It was not collected in the Northwest until after 1890, when it was located in Montana, Idaho, and Washington. It spread to Oregon by 1910 and to Wyoming by 1940.[113] Unlike other range plants introduced about the same time (downy brome, for example), yellow toadflax was apparently not abundant or widely distributed enough to warrant inclusion in the *Range Plant Handbook* in 1937, although contemporary weed specialists indicate that since then it has become "an aggressive invader of rangelands, displacing desirable grasses."[114]

Of course, if western land had not become valuable (and overused) as a source of livestock forage and toadflax had remained a useful and even cultivated plant, it is doubtful that it would have had occasion to become an undesirable invader on the western range at all. Kochia was another ornamental plant that escaped in the West, becoming common in dry western fields and pastures by 1911.[115] Unlike yellow toadflax, however, it did not have a long history as a multiply useful (as well as beautiful) plant; its original value is based simply on the fact that it turns bright red in the fall. But kochia has one other important virtue: "While it is usually considered an objectionable weed, kochia is readily grazed by livestock."[116] Above all else, whatever a plant's significance might have been before it came to the United States, once here, it was ranked by the degree to which it either was palatable to livestock or did not interfere with the cultivation of other plants.

For western farmers and ranchers, as for farmers and herdsmen everywhere, weeds were certain to be part of any agricultural endeavor. The conviction that weeds should be excluded from agriculture, however, led to the use of specific strategies for maintaining monocultural fields and uniformly palatable pastures. One such strategy was passage of antiweed legislation, whereby weed seeds were outlawed in commercial crop seed and weed plants were banned from fields and rangeland.

Western states often passed weed legislation relatively early in their legislative lives. Connecticut passed the nation's first weed law, banning the sale of forage grass seed containing Canada thistle seed, in 1821.[117] Nebraska passed its first weed law in 1873—only six years after being admitted as a state and two years before ratifying the state constitution. Nebraska banned both the sale of grass seed containing Canada thistle seed and Canada thistle plants themselves. Landowners were obliged to cut or mow any Canada thistle growing on their property before it went to seed, and anyone who entered the property to cut or mow the plants could not be sued for trespass.[118] Weed legislation began in South Dakota with the first session of the legislature in 1890, and Kansas likewise began legislating against weeds at the end of the nineteenth century.[119] The federal government prohibited the interstate transportation of weed seeds in commercial seed and grain in 1938 (there was no federal regulation of weed plants until 1974). By that time, all of the western states had enacted weed legislation of their own as well.[120]

Weed seeds had never been desirable in crop seed or grain, but they presented a new set of problems to the industrialization of grain and flour production that had begun with the western wheat boom. Grain elevator owners routinely docked wheat if it contained weed seeds that their machinery could not remove.[121] Weeds in the field made extra work for draft horses and caused more wear on farm machinery, sometimes making hand-weeding or hand-harvesting necessary, which was "the most expensive form of labor in every occupation."[122] The industrialization of harvesting, threshing, and milling wheat transformed the fields into sprawling wheat factories and weeds into obstacles in the smooth production of commercial grain. Without the demand for large-scale, mechanized grain production, weeds would be no less present in grainfields, but they could be dealt with on a scale small enough that hand-pulling and hand-harvesting were not unthinkably inefficient and did not represent a failure of the grain-growing process.

For centuries, European and Euro-American farmers' response to weeds was preventive (attempting to keep weed seeds out of the seed grain) and mechanical. Farmers everywhere have used their hands, sticks, and hoes to remove unwanted plants. In the eighteenth century, Jethro Tull launched modern European weed control when he created the modern field planted in long, straight rows. If you planted crops in appropriately spaced rows, he determined, you could hoe the fields with horses pulling a specialized tool. Not accidentally, he was among the first to use the word "weed" with its present spelling and meaning.[123]

The displacement of the farmer's hoe by the horse-drawn cultivator was an early move in the direction of "labor-saving" machines for the field, maximizing the productivity of the farmer's labor at all stages of agricultural production. The horse-drawn cultivator also established the modern weed as something to be removed from the field with maximum efficiency. Cultivation remained the primary method of weed control into the 1940s.[124] For annual weeds, it was important to prevent the formation of seeds; for perennials, cultivation would eventually starve the roots by repeatedly destroying the new leaves. Other methods of mechanical weed control included mowing, plowing, harrowing, and disking, all performed by different tools drawn by horses or later by tractors.[125] One could always resort to hand-hoeing (or even pulling by hand), but "this method is too laborious except for small areas. It is useful in small gardens and in rows of cultivated crops and for supplementing machine cultivation late in the season when the crop plants are large enough or to destroy scattered weeds missed by the cultivator."[126]

Mechanical weed control went hand in hand with the increasing cultivation of large fields and demand for efficiency. Techniques viable in gardens or smaller eastern fields were inconsistent with the scale and purpose of agriculture developed in the West, which eventually shaped the expectations of agricultural production for the whole country.

Rangeland presented a weed control problem specific to the West primarily because of the amount of land affected by weeds after the livestock booms and agricultural settlement. Mechanical weed control was out of the question in most cases, and chemical control, once it became available, was too expensive, sometimes actually exceeding the cash value of the land.[127] Throughout the twentieth century, then, range weed control has taken place, just as range agriculture, by the careful management of livestock grazing to promote the growth of desirable species and interfere with or prevent the growth of less desirable plants.[128] As British botanist Winifred Brenchley noted in 1920, "The consideration of the weeds of grass-land presents a very different problem from that of arable weeds because of the totally different nature of the crop with which the weeds are associated." All plants sown in a field are "alien" to the bare soil, whereas on the grassland "the crop consists of an association of plants," and the only criteria for judging the weeds is their palatability and whether or not they are poisonous to livestock.[129]

The farmer's utilitarian relationship with the range did not change merely because the crop in most places was not planted in cultivated ground. The indirect domestication of the range embodied an indirect means of weed control where no other form was possible. One might rejoice that chemicals have continued to be expensive enough to discourage their indiscriminate use on the western range, but the significance of range weed control remains. Had livestock not become such big business that cattle and sheep were herded onto the grasslands and into the mountains by the tens of millions, the question of weed "invasions" (like that described by Richard Mack regarding downy brome) would not have come up. Had the range not been "agriculturalized," maximizing the productivity of flesh and grass, weeds would not have been identified as interfering with range productivity.

By the twentieth century, chemists and weed specialists had turned a corner in the development of efficient weed control, adding chemical herbicides to the available strategies of weed prevention and removal in cultivated fields. The development of herbicides coincided with the development of the science of chemistry, which began to discover many agricultural applications for chemicals by the end of the nineteenth century. Plant nutrition had been described in chemical terms before 1870 by German chemist Justus Liebig, who developed chemical fertilizers to augment agricultural plants' supply of nutrients.[130] The development of chemicals to stop the growth of unwanted plants was a corollary to this emerging interest and expertise in the chemical activity of plants.

One of the first researchers reporting success with new herbicidal chemicals was an American, H. L. Bolley, who was studying weeds in North Dakota grainfields. When he published his results in 1908, he had already completed twelve years of experiments with common salt, iron sulphate, copper sulphate, and sodium arsenate, which he used to kill "broadleaved weeds" (anything that was not grasslike) in cereal fields. Chemicals that can be used in this way, killing the weeds and sparing the grain, are known as selective herbicides.

People had been using chemicals such as salt, ashes, and industrial by-products for a long time to kill plants indiscriminately where they were not wanted. But selective herbicides were new, and Bolley understood that his success would have great economic significance: "When the farming public has accepted this method of attacking weeds as a regular farm operation, . . . the gain to the country at large will be much greater in monetary consideration than that which has been afforded by any other single piece of investigation applied to field work in agriculture."[131] Bolley's colleagues in Wisconsin, Iowa, and South Dakota reported similar results in 1909 and 1910. Six out of eight researchers developing chemical control of weeds in grainfields at that time were working in the West or the Midwest, underscoring the significance of the western grain industry, and those scientists entrusted with its advancement, in defining the conditions of twentieth-century agricultural production.[132]

The development of new chemicals and chemical applications in the West and elsewhere in the United States was slow after 1910. Early herbicides worked well only in areas of high humidity, and most American grain production occurred in dry places. Western grain farms were typically very large, and spray machinery had not yet been developed to accommodate the scale of the fields to be treated. Most of the weeds subject to early forms of chemical control were annual weeds, and in the absence of efficient chemical control, other methods were developed to deal with them, including increased attention to weed-free crop seed, as we have seen. By the 1920s, interest in chemical weed control was shifting to the control of perennial weeds and various methods of soil sterilization or the complete and permanent killing of vegetation. Arsenic-based chemicals proved satisfactory for these purposes and were widely used in the West, not in the fields, but along railroads, highways, and firebreaks.

Industrial waste also found uses as soil sterilizers, including iron sulphate produced in the manufacture of wire by the American Steel and Wire Company of Chicago, whose executives no doubt hoped to cash in on the unexpected value of their by-products at the same time that they made their fortune fencing the West. Other industrial wastes used as weed killers included sulfuric acid from petroleum refineries and mineral smelting, creosote, used crankcase oil (sometimes diluted with kerosene), and unrefined petroleum products generally. Any of these chemicals could be used to kill small stands of perennial weeds where they grew.[133] Agriculture was certainly "following the lead of industry," as Wilfred Robbins and his colleagues wrote, by "resorting to chemical aids," which included the toxic leftovers of industry as well as chemicals developed specifically for agricultural purposes. The need for new chemicals was particularly acute in the West. By the 1930s, chemical methods had "received increasing attention, particularly in the newer districts, where large-scale crop production by machine methods is found" and efficiency was paramount in making a profit.[134]

A French herbicide developed in 1933 signaled a new era in weed killing at the same time that it continued the relationship between industrial waste and herbicides. Sodium dinitro-orthocresylate was marketed in the United States as Sinox after 1937. It was used as the dye "victoria yellow" and was made from by-products of coal distillation. In the late 1930s, little was known about how Sinox actually worked physiologically, but it was clear that the chemical did not simply burn or kill plants through contact alone, as so many other soil sterilants did. Plants sprayed with Sinox appeared to over-respirate, indicating a chemical intervention in the plants' metabolism more sophisticated than the caustic tissue damage caused by older herbicides. The chemical was tested in Oregon and California, applied from the air by planes, and found satisfactory in controlling weeds of grainfields and vegetable fields without harming either grain or alfalfa. Sinox was known to be both "extremely toxic and highly selective," excellent qualities for a herbicide. It was noncorrosive and relatively nonpoisonous compared with sulfuric acid and arsenic-based herbicides. Sinox was organically broken down in the soil without leaving a toxic residue.

Sinox seemed to incorporate "the advantages of all other selective herbicides, with an almost complete absence of their drawbacks, and makes an ideal chemical for weed-control purposes." Robbins and his colleagues predicted that "if present experiments with this and other similar organic compounds can be taken as an index, it seems possible that within the not too distant future there may be available a number of selective herbicides for use against different groups of weeds and that the differential killing of noxious weeds in pastures, turf, and grainfields may become much more commonplace and reliable than it has been in the past."[135]

Advances in plant chemistry and physiology led to the development of a synthetic plant growth regulator, 2,4-dichlorophenoxyacetic acid (known as 2,4-D), in 1941, which fulfilled Robbins's expectations. This chemical was made to mimic plant hormones that stimulated the growth of tissue and increased the rate of plant metabolism. Two thousandths of an ounce of this chemical, when mixed with lanolin and applied to the stem of a bean plant, resulted in the growth of roots along the stem that functioned just like roots the plant developed on its own if covered with soil. In addition, 2,4-D could be used to make holly produce berries from unpollinated flowers or to stimulate root growth in greenhouse tomatoes. If sprayed on plants' leaves, 2,4-D caused the plants to stop growing, use up their stored food supply in their stems and roots, and "starve to death."[136] Relatively inexpensive and easy to apply, 2,4-D was eventually used to treat more acres of small grains than any other crop treated with any other herbicide.[137] Because it affected an entire plant's metabolism, rather than merely killing its aboveground parts, 2,4-D could be used on perennial weeds like Canada thistle and Russian knapweed. With enough moisture, 2,4-D would remain toxic to plants in the soil for less than a month, allowing desired crops to grow soon after application.[138]

Within a few years of its development, 2,4-D had already become "one of the most successful weed killers we know."[139] The success of 2,4-D as a herbicide and the continued development of chemicals and chemical techniques eventually changed the terms of weed control in the West and elsewhere. Robbins and his colleagues still advocated controlling weeds by tillage in 1942 and appealed to farmers' own skill as the ultimate defense against weeds: "Other methods of weed control, such as the use of chemicals, may have a place, but their employment is many times a poor substitute for tillage and cropping and represents a mistaken belief on the part of the grower that there is something magical in the behavior of herbicides. It is literally true that the grower who consistently follows good farming practices [rotating crops, using weed-free seed, cultivating assiduously] has little trouble with weeds."[140]

But by 1947, 2,4-D was in use on a large scale. By 1957, 92 percent of all farmers depended regularly on herbicides, and weed science had turned away from plant identification and the history of weed introductions, more common among turn-of-the-century botanists, to emphasize the types and proper use of chemical herbicides.[141] A 1951 USDA Agricultural Research Administration bibliography of weed investigations devoted about ten pages to citations regarding chemical control of weeds to every one page on works addressing mechanical weed control.[142]

Although farmers might easily have adopted cheap, "efficient" chemical weed control on their own, the role of George W. Merck (president of a large chemical company bearing his name) as head of the new War Research Service in 1942 ensured the production and legitimation of chemical herbicides that could be marketed in peacetime to farmers.[143] E. J. Kraus, head of the Botany Department of the University of Chicago, suggested using the new synthetic growth regulators like 2,4-D as plant-killing weapons during the war. Kraus and his colleague J. W. Mitchell completed tests on these chemicals at the University of Chicago, receiving a contract from the U.S. Army in 1943 for the work they had already done. Based on Kraus and Mitchell's report, the War Department took up herbicide research at Camp Detrick in Maryland, where over a thousand defoliant herbicides were tested before the end of the war.

The purpose of these chemicals, of course, was still "weed control": in this case, the destruction of the enemy's crop plants, which in time of war were undesirable and threatening precisely because of their use as sustenance for human beings. Herbicide weapons were not used against the Germans or Japanese (although this type of research led directly to the herbicidal compounds used to such destructive and toxic effect in Vietnam), but they became an arsenal against weeds at home in a rapid transfer of military technology to civilian production and consumption.[144]

Herbicide research in the United States had begun with the need for weed control in vast western grainfields; it culminated in the development of herbicides as weapons and contributed to the postwar entanglement of military and corporate interests. Herbicide research also maintained the connection between plants and their people, targeting one in order to destroy the other during wartime.

The idea that a herbicide could have both military and agricultural significance is built into the concept of the weed, especially as it developed in American agriculture. Indeed, the military application of herbicides was all but guaranteed by the convention of excluding and destroying those plants understood to threaten agricultural productivity. This convention had always had social implications. In the imagination of the botanist Darlington, invading plants overtook native vegetation just as Native American populations "disappeared" following the invasion of Europeans. For ecologist Frederic Clements, people were subject to succession just as plants were, progressing from the Maya through a series of invasions by the "Toltec, Aztec, and the Spaniard in Mexico, by the various Pueblan cultures of the Southwest, and by the trapper, hunter, pioneer, homesteader, and urbanite in the Middle West."[145]

All of Clements's ecological categories were assigned names and functions with profound social significance. The "movement of plants from one area to another, and their colonization in the latter," was an invasion. A colony was a "group of two or more species that develops in a barren area, or in a community as an immediate consequence of invasion." On land disturbed by climate or human activity, annual "pioneers" were the first plants to appear, and they "regularly disappear in the course of development." Weeds were among these pioneers, some of them "subruderals," or native plants that invaded barren areas, and the rest were "ruderals," or introduced plants growing under disturbed conditions.[146]

Clements's association of native plants with introduced plants—all of them ruderals, or weeds—helps explain how European plants could be as undesirable and likely to disappear as Native Americans: both were part of the most primitive phase of colonization. Native populations had no meaning within a colonial ideology beyond their obligation to vanish in the face of progress, understood first as agricultural and then as industrial; trappers, traders, and other "pioneers" were primitive representatives of the settlement to come. In the West, both weeds and Native Americans became the objects of increasingly "efficient" projects of removal.

Before weed control was turned to the service of the military, the military had embarked on a systematic campaign of Indian removal throughout the West after the Civil War, culminating in the massacre of Dakota people at Wounded Knee in the winter of 1890. Like weeds, Native Americans could be accused of aggressive behavior simply by their refusal to be eradicated.[147] They claimed ground suited to agriculture, a form of occupation "higher" even than their own food production, and were expected to give way ultimately by force to colonization. It is inconceivable that Clements could have imagined weeds as pioneers in a theory of "succession" if anti-Indian ideology and military persecution had not already become entrenched as forms of human-weed control.

People had perhaps always been the first weeds, from the dispossession of sixteenth-century farmers, to the exclusion of women's plants, techniques, and expertise from field agriculture, to the removal of Native Americans encumbering western ground thought to be "ready" for the plow or the hoof—all beginning points in stories about agricultural progress, as we have seen. In this context, it is a short conceptual leap from battling people to battling weeds and back again. Should we be surprised that the insecticide DDT was advertised as the "Atomic Bomb to the Insect World" in September 1945, less than a month after the bombing of Hiroshima and Nagasaki?[148]

What is remarkable about all weeds, human and plant alike, is their persistence in the face of colonization, mechanical and chemical wars, systematic exclusion, and policies of eradication. There is no evidence to suggest that the acreage of weeds or crop losses to weeds have permanently decreased since the advent of 2,4-D.[149] Native American spiritual, cultural, and political activities are vital parts of contemporary society in the United States, in spite of the racism, warfare, disease, compulsory assimilation, and land loss that Native American communities have endured for centuries. There are, of course, many other human weeds as well. Instead of being a sign of an agri/cultural failure, the persistence of weeds despite weed laws, cultivators, and 2,4-D should give us hope that the commercializing, homogenizing forces of agriculture will have to give way to more complicated fields and societies.

Thorns and thistles have long been scratching at the margins of agriculture. They have also inspired courage and tenacity in people marginalized by the racial, cultural, or economic imperatives of western agriculture, as one final plant history illustrates. Alfilaria (*Erodium cicutarium*, anglicized as filaree) was introduced in the American Southwest by the Spanish as a forage crop.[150] Like alfalfa, alfilaria's name is Arabic, and this plant was probably among the crops introduced in Spain by North African colonizers. Unlike alfalfa, however, alfilaria never became a staple forage plant of western Anglo ranchers.

By the 1840s, alfilaria had "taken possession of large tracts" in California and Oregon.[151] In 1911 there was a "constant danger" of the introduction of alfilaria into new areas in wool from western states, as the plant was common in "much of the western country" and abundant in the Salt Lake basin.[152] The *Range Plant Handbook*, and presumably many ranchers in the West, recognized the value of alfilaria as forage at the same time that it called the plant an "aggressive invader."[153] Alfilaria spread from California into Arizona with the large influx of sheep into that state in the 1870s.

It was there that the hard-struggling, working-class protagonist of Marguerite Noble's novel *Filaree*—modeled after the novelist's mother—pondered the relationship between the vitality of filaree and human vitality and courage:

> In drouth times, the land battled the aridity. Some plants died, and others adapted, like the cactus. Some resisted and survived—like the filaree. The nerves and spirits of The Mesa people became as tattered and frayed as the desiccated plants. They swore at the land, and some fulfilled their threat to move on—or go back to where they came from. But when the rains came, the woman had seen her husband ride down the draw with the filaree belly-high to his horse, and his stirrups brushing the mariposa lilies. In such years, The Mesa folk recovered and forgot. . . . In drouth, the plant hugged the earth and waited—and survived in its tenacity. She thought of herself in the same way: waiting for the rain, waiting for life to give her something—something to fulfill her destiny.[154]

In this novel, the heroine's longing is directed against an endless condition of pregnancy, migraines caused by the heavy long hair her husband forbids her to cut (or perhaps by the injunction itself), and her general subjection to her husband's abuse and the economic hard times she is forced to weather on his terms. Hers is not a "manifest destiny" but an obscure one endangered by the imposition of structural and arbitrary power over her life at the expense of her own agency and vitality. This character's use of filaree is no simple parable of a Darwinian struggle "with the land" or "with aridity." She struggles against patriarchy and poverty, conditions perhaps made more difficult by a drought but formidable even in the best of seasons, conditions shot through with physicality as well as social hierarchy, a state of affairs aptly embodied by the filaree. Here is a plant with colonial value to North Africans and Spaniards, escaped and become weedy with other social meanings to the Anglos, recuperated again in a new way by a poor pregnant woman in Arizona in 1910.

This is precisely the nature of weediness—to get away, to spread both outside and inside the field of cultivation and domestication, to be variously absorbed into social orders as an outlaw invader, a model of tenacity, a minor nuisance, a beautiful plant, a poison, a crop, a holy or cursed thing. Nothing about the weed itself determines which of these things it can be. It is neither good nor bad, but, of course, potentially both. Agriculture, as such, is fundamentally about domesticating a category of "not-weeds" from which everything else is formally excluded. The terms "weed" and "not-weed" misleadingly suggest that there is such a thing as a weed from which many not-weeds can be distinguished or that the variety of weeds is unimportant. In fact, the opposite is true. Not-weeds represent what is selected, isolated, and privileged to the exclusion of an amazing variety of weeds, whose threat lies in their capacity to overwhelm the desired efficiency of agriculture, especially agriculture as it developed in the American West.

This capacity is not genetic, in spite of the identification of allegedly weedy characteristics. Such an identification only reveals the characteristics of agriculture by defining in the negative those properties in plants that it cannot tolerate and against which it must constantly defend itself. An agricultural society suffers from a curious myopia in this way. Not only does its agriculture exclude literally everything and everybody whose "productivity" cannot be exploited and maximized—a narrowness of both social and material vision—but when it looks outside itself, the unintelligible variety of what lies outside becomes a mirror. The aggressive colonizers look at the thistles for which their fields are open invitations and see aggressive colonization. The ecologist trained in sciences founded on the idea of progress—of history, knowledge, social, and biological complexity—looks at many plants in many settings and sees orderly progression, indeed, colonization.

This state of affairs illustrates more than simple ethnocentrism, whereby a society can only interpret the world through its own limited reference. An agricultural society is founded not just on its difference from other societies but on the exclusion and extirpation of other societies. As it pushes out over its territory with its projects of colonization, occupation, and the formation of the advanced state, whatever pushes back—plants, insects, Native Americans, unwanted immigrants—looks, to the colonizers, like themselves. Weeds become a horrifying doppelgänger, bearing an uncanny resemblance to the colonists who exclude them, if only because the colonists are so ill-equipped to understand weeds in any other way. It is important to note that this uncanny resemblance between the invading wheatfield and the invading thistle lies wholly within the society of the wheatfield, which is not itself a "weed" but an invading state. The alleged "aggressiveness" of the weed only mirrors the desire of agriculture for complete domestication of its territory.

What is more interesting than the weed's agricultural ingress is its egress—its perpetual getting away. An agricultural state represents an imposition of rule, at once social, biological, epistemological, and technical. A weed represents what defies or escapes that rule. Gilles Deleuze and Félix Guattari use several images to describe a distinction of this kind: between smooth space and striated space, between the nomad and the state, between lines of flight and closed circuits. Each of these images was played out in western agriculture, as the "unbroken" prairie received the striations of the plow, as the hunters and gatherers were confined to the state's reservations, as human and animal movement was increasingly limited to permanent routes tying cities and settlements to one another. The most apt image for this distinction, in the context of weed control, is the rhizome, used by Deleuze and Guattari as an image of something that gets away from the permanent imposition of order—those things that are "arborescent," hierarchical growths that impose unity on the multiplicity of things.

The rhizome is in part a metaphor for all things that have multiple roots instead of single roots that define an overarching meaning or inform the structure of a thing. The rhizome is a metaphor for the multiple itself. "Rats are rhizomes," "in their pack form."[155] Agriculture is an arborescent structure whose single "root" is cultivation. All of its multiplicity is buried underground, completely integrated into the organization that is the growing "tree," the flowering of civilization. Agriculture relentlessly transforms nature into culture. An agricultural society minutely examines the world to define its parts, harness their energy and metabolism, determine their purpose—its overdeveloped eyes and ears like those of the wolf who eats the grandmother, all the better to eat you with, my dear.

The rhizome, on the other hand, escapes the single tree (and the "singletree," the harness); it is always multiple, resourceful, springing up everywhere. The rhizome is not just a metaphor; it also describes actual plant parts, "bulbs and tubers." Whether it is rats or roots, though, metaphoric or descriptive, the tendency of the rhizome is to be many things. "The rhizome includes the best and the worst: potato and couchgrass, or the weed." This is exactly the ambiguity of weediness fundamental to weed identification. Moreover, not only is the "good" mixed up with the "bad," but whole ontological categories become confused: "Animal and plant, couchgrass is crabgrass."[156] The rhizome is where the distinction of parts and their hierarchical assemblage into an organism breaks down, whether that organism is a climax community of grasses or a progression of "pioneers" on the western landscape culminating in the advanced agricultural state.

As plants that get away, weeds are instructive of how narrow and instrumental the relationship is between agriculture and plants. As examples of rhizomes, weeds illustrate the promiscuity of plants and people and the possibility of an existence outside agri/culture. They remind us that if thistles are invaders and filaree is a patient woman, people can escape the agricultural order that feeds them and domesticates them. Weeds overrun and undermine the whole project.

Epilogue. Just the Facts, Ma'am

> The question is not: is it true? But: does it work? What new thoughts does it make possible to think?—Brian Massumi, translator's foreword to Gilles Deleuze and Félix Guattari, *A Thousand Plateaus: Capitalism and Schizophrenia*

A hundred years ago, Frederick Jackson Turner offered an interpretation of the American frontier—newly "closed"—that commanded a great deal of attention during the first few decades of professional history in this country. He argued that the frontier was "the line of most rapid and effective Americanization" of onetime Europeans in the "new world." The drama that unfolded at this line, resulting in an entirely new nation, was no less than a great adventure:

> The wilderness masters the colonist. It finds him a European in dress, industries, tools, modes of travel, and thought. It takes him from the railroad car and puts him in the birch canoe. It strips off the garments of civilization and arrays him in the hunting shirt and the moccasin. It puts him in the log cabin of the Cherokee and Iroquois and runs an Indian palisade around him. Before long he has gone to planting Indian corn and plowing with a sharp stick; he shouts the war cry and takes the scalp in orthodox Indian fashion. In short, at the frontier the environment is at first too strong for the man. He must accept

> the conditions which it furnishes, or perish, and so he fits himself into Indian clearings and follows the Indian trails. Little by little he transforms the wilderness, but the outcome is not the old Europe, not simply the development of Germanic germs, any more than the first phenomenon was a case of reversion to the Germanic mark. The fact is, that here is a new product that is American.[1]

The "he" of this epic is supposed to represent all of the real individuals who underwent this transformation at the hands of the wilderness, but as a narrative device, his singularity is telling. He draws attention to the fact that Turner is narrating, that what concerns Turner in this thesis is the outcome of a story.

Turner's narrative concern turns to the continent itself when he notes that the "United States lies like a huge page in the history of society. Line by line as we read this continental page from West to East we find the record of social evolution."[2] In the East, the text becomes dense, as "this page is a palimpsest." The drama of the wilderness was legible where it was most recent (on the frontier), culminating in the East where the evolution of society had overwritten its frontier past in its evolutionary development. Writing this story was a quintessentially American act for the individual, wherever he happened to reenact it, and for the nation as a whole, across the broad page of the continent. Now that the frontier had closed, the story could not be started anywhere else, leaving the page of the nation to fill in with palimpsests. "There is not *tabula rasa*," Turner wrote.[3] The closing of the frontier was a narrative fact as well as a demographical one.

In the waves of criticism, qualification, attack, and defense that followed, Turner's commentators measured his thesis against "the facts," and many found the thesis and the theorist wanting. Turner never wrote a book-length work, and in the essays he published (and republished), his thesis was "not proved so much as it [was] continually reiterated" by himself and later by his students and defenders.[4] He was provincial, he exaggerated, he made sweeping generalizations; he neglected the development of the cities, the effects of immigration, and the rise of monopoly capital on American institutions. At worst, "Turner and his followers were the fabricators of a tradition which is not only fictitious but also to a very large degree positively harmful."[5] Its harm lay, in part, in not inspiring students to be responsible historians. They repeated Turner's ideas without subjecting them "either to the analysis or to the tests which the rudiments of scientific knowledge would seem to suggest,"[6] and the profession of history had become very serious about its scientific credentials. In Turner's own generation, "neither Turner nor his contemporaries were aware of the fact that he was violating the canons of scholarship" by writing "with the undisciplined impetuosity of the poet rather than with the restrained exactness of the scientist."[7]

If Turner's fault was that he "viewed history as an art rather than as a science,"[8] even his critics acknowledged his skill. He was "a poet in the grand manner" and in his "poetic capacity [wrote] brilliant and moving odes to the glories of the westward movement." His "romantic passages" made for a "splendid piece of writing."[9] He had not, however, intended to write "merely" fictive romances. He wrote to Merle Curti, "I always wanted to be an artist, tho' a truthful one."[10] The significance of Turner's work as historical *writing*—including Turner's conflation of the American continent with the pages written about it—was lost on his critics.

Because they were not convinced by his writing, Turner's critics were not concerned with identifying the genre in which he wrote. He was, in fact, an explorer, not of a real "wilderness" but of the textual historiographical one, where continents can become legible pages—where continents must become legible pages after there are no more wildernesses to conquer. Turner wrote of his work that he had " 'had a lot of fun exploring, getting lost and getting back, and telling my companions about it.' "[11] He explored the idea of the frontier as a historiographer, taking notes in the underappreciated backcountry of western history, reading the settlement of the continent like a book, and offering his well-composed traveler's diary to an apparently eager audience. When he wrote about the frontier, he himself became the protagonist of the western epic, a role he moved in and out of as he addressed other interests: "I don't quite care to figure in leggins and breech clout all the time," he wrote.[12] He explored the frontier of American history this way because it was no longer possible to be an explorer "for real." He was an earnest backcountry loiterer after the model still current—a trout fisherman who spent his summers in the West or at his house in the woods of Maine.[13] He had read about the explorers and studied the early papers of Wisconsin fur traders, but his experience as an explorer was vicarious and textual. One of his favorite poems was Rudyard Kipling's "The Explorer."[14] Turner's thesis was an exercise in nostalgia for the explorers who had gone before, leaving him only the possibility of a historian's romance with the blank page, on which he could turn a fine phrase, and a good time fishing in well-mapped territory. Francis Parkman died in 1893, and with him died the possibility of chronicling the newness of the Great West. It was this narrative frontier that closed in the last decade of the nineteenth century, and Turner wrote an obituary.

Turner's essay was not the only expression of a narrative concern with a crisis in imperialism among colonial writers at the end of the nineteenth century. The "endless stream of late-Victorian explorers' narratives, campaign reports, and missionaries' tales" ran headlong into "the moment of 'late imperialism,' when expansion reaches its global limits and resistance to domination becomes increasingly articulate."[15] In the United States, where 1890 marked the end of large-scale warfare with indigenous peoples (after the massacre at Wounded Knee) as well as a low point of Native American self-determination as mandatory allotment swept through reservations, the competition for limited space was felt by Euro-Americans more in regard to immigrants from Europe than in regard to native resistance. But the collision with a limited globe was no less dire for that. Writing as an American in favor of immigration restrictions, Josiah Strong commented in 1885 on a more general situation: "There are no more new worlds. The unoccupied arable lands of the earth are limited, and will soon be taken."[16]

This collision with limits produced a new kind of romance among writers like Joseph Conrad and E. M. Forster, a more or less critical look at the previous generation's "glories" of colonial expansion, with a longing all the same for the time before all the good stories had been told. Turner was less, rather than more, critical of what had gone before but was clearly preoccupied with the issue of telling the story about it as a narrative explorer of the American romance. In the case of Conrad, the blank spaces on the map that his narrator Marlow had wished to visit (and which were eventually filled in) were far away—Africa or the North Pole, fabulous places to a European imagination. For Turner, who grew up in Wisconsin, the blank spaces had been quite close, had indeed relatively recently included his birthplace. The narrative of colonial adventure, which in America became a story about freedom and opportunity in "new lands," had been used to describe what took place in Turner's backyard. It is no wonder, then, that the end of a version of that story would present itself to Turner as an important punctuation in American history and in his own life.

The closure of one frontier—the acknowledgment that the world really contained no more blank spaces—immediately opened a new one for Turner in the territory of western historiography. But for him, at least, that frontier was as limited as the geographical one. Without a rhetorical structure that allowed something new to happen after the close of the frontier (since that alone distinguished America from Europe), and as long as Europe represented the antidemocratic forces of capitalism and privilege, a sustained story of the triumph of American democracy was impossible for Turner. Other historians, for example, Charles Beard, overcame this narrative difficulty by seeing in European industrialism the kind of opportunity for democracy Turner had thought unique to the American West.[17] So the narrative frontier (and the idea of American opportunity) remained open, without an examination of the issue of limits, let alone the legitimacy of the story of colonial conquest in the first place, that Turner's narrative preoccupation and difficulty might have begun. Critics and defenders missed an opportunity, or avoided the necessity, of taking an imperial story and its expiration seriously.

Regardless of the complaints of historical realists, science was and remains an act of composition, identifying and organizing what will come to be naturalized as "facts." One of the most significant facts at Turner's disposal was the data of the 1890 census, and his use of it identifies him as a subscriber to the truths of science, even if his writing identified him otherwise to his readers. A census is a kind of snapshot of the nation—the number of its people, the value of its resources—but it is itself an artifact, evidence of the kind of nation that would find such tallies valuable.

Thomas Richards describes the importance of tallies and surveys, and collections of data in general, to the administration of the British Empire.[18] British colonial administrators really had no hope of managing the various territories of the empire with any absolute power. What they created instead was the illusion of power through the accumulation of data from the colonies. Explorers, surveyors, scientists, military administrators, all collected data and added it to an imperial archive whose comprehensive grasp of information substituted for the limitation of power these colonials actually wielded. Richards finds the fantasy of comprehensive knowledge, as embodied in the imperial archive, expressed repeatedly in British literature, from the assemblage of information that identifies and defeats Dracula to the transformation of the sea into an underwater museum in the world of Captain Nemo. At the very moment that a historian taps the imperial archive for a pure draft of the truth, what he gets is the cumulative fiction of colonial knowledge gathering and control. To charge Turner with failing as a scientific historian then puts the critic in an absurd position.

While historians believed their work was becoming more and more scientific—that is, "truthful," based increasingly on the facts of the past as elaborated by census takers, public reformers, autobiographers, anthropological subjects, archaeologists, geographers, ecologists—scientists themselves never gave up the historical element of their own work. The stories told by agricultural scientists, for example, serve to organize and give meaning to the data they collected in the field, justifying their work and placing it in a progression toward ever more efficient and productive agriculture. The documents of the USDA, like the census, are very much an imperial archive. The USDA has been a tremendous knowledge-gathering institution; from the beginning, it kept records of its work, published in bulletins, circulars, booklets, and annual reports, all of which are shelved in the library of every land-grant institution in the country. Unlike other archives, where papers are gathered after somebody's death or collected from many places and types of sources to illustrate a specific historical problem or one archivist's obsession, the USDA archive is an ongoing project of thousands of people, like the imperial archive Richards describes, gathering data that represent increased agricultural productivity and profit, the signs of success for colonial agriculture.

The story agricultural scientists tell countless times, often explicitly, is an epic of civilization that places themselves at the apex of development. The orderliness of this archive cannot be disputed, and every bureau in the whole machinery leaves an exemplary paper trail. To enter this archive is like entering the Heart of Darkness, like Marlow, who noted that conquest is not a pretty thing when you look into it too much. What the USDA has documented since 1862 is the systematic conquest of the wild or the natural, and its permanent replacement with the tame and the useful. But also like Marlow, one can be fooled into believing that what lies in the archive is something logical because it is systematic. The progress of civilization runs through the work of weed scientists and animal scientists, dendrologists and agrostologists, and all sorts of other specialists, from the 1860s more or less through the present. The archival explorer can follow plants and animals through indexes and tables of contents and the careers of individual researchers. USDA material is nothing if not well marked. But the sheer volume of information belies its order.

The archive contains information about every conceivable aspect of agriculture—circulars on chicken yard sanitation, marketing steers, building a barn, choosing varieties of wheat, locating a ranch, irrigating orchards or alfalfa or pastures. If anyone really knew these things, no farmer would go out of business; the land would not blow away in a drought or wash away in great sheets and gullies; animals would live in clean pastures; people's vegetable gardens would be bountiful, ubiquitous, large, tended with care and confidence. Charles Greathouse published a bulletin on the farmer's vegetable garden in 1899. He described the layout and content of the garden and its care, noting that it should be close to the kitchen. W. H. Beal wrote a circular in 1902 of exactly the same type. W. R. Beattie wrote a garden bulletin in 1906 including almost identical information. Then Beattie published again in 1931, this time with J. H. Beattie, what was clearly only a revision of the 1906 article on the farmer's kitchen garden. Such iterations do not represent "facts" of gardening so much as they appear to represent the imperative to write authoritatively about gardening.

The mass of USDA material is beautifully ordered, but as Marlow discovered about Kurtz, the archival traveler finds that the USDA had no method. What the USDA has actually produced all these years is not agricultural progress: it has produced *itself* through its immense archive and its self-reflective and congratulatory histories. The staid little yearbooks of the late nineteenth century gave way to flurries of bureau and division reports and innumerable circulars, bulletins, and pamphlets by the 1920s. This proliferation of agricultural expertise, sent diligently into the archive through the Government Printing Office, is a monument to perhaps the greatest agricultural de-evolution known to human societies. Like the imperial archive that Richards locates in British fiction, the USDA archive represents the fantasy of empire, the unrealizable goal of comprehensive control and uninterrupted progress, at the same time that it documents the work of agriculture in bringing culture to a wilderness.

Whatever Turner's failures as a scientific historian are, they should be viewed within the context of the ambition and limitation of the imperial archive from which he drew when he cited the 1890 census and that ultimately included many repetitions of the frontier story in the rigorously scientific discourse of agriculture. To draw attention in this way to the epistemological groundlessness of agricultural (and every other) science has become a kind of postmodern parlor trick, but my point here is not theatrical. The so-called crisis of representation is perhaps only a crisis within the imperial archive, the intellectual consciousness of the limits of knowledge and ultimately power.

To recognize Turner's work as relying on a fiction widely held by scientists and nonscientists alike is not to throw human experience into the teeth of nihilism but to begin a process of reclaiming knowledge from the archive. The repatriation of Native American remains and artifacts long held in museum collections would be an example of such reclaimed knowledge. What was "data" to the museum becomes something else entirely to the people to whom it belongs. In agricultural work, the imperial archive continues to absorb information about seeds and crops, and now germ plasm itself, in order to patent organisms as "products" and sell them back to farmers. Counterarchivists like Gary Nabhan are also at work, reclaiming and restoring seeds threatened by the centripetal force of specimen collection, classification, and "improvement."[19] The counterarchival fields of Native American beans or corn are not museums but living repositories of both social and agricultural memory that belong to the people who plant them, and they exceed in every way a narrow identification of a plant as a "crop." Indeed, the agricultures Nabhan writes about and works with show that the distinction between the wild and the cultivated is wholly inadequate to describe what happens between people and their plants when left to themselves outside agri/culture as such. Every aspect of the imperial archive of American agriculture must be reclaimed in the same way in order to interrupt the manufacture of agricultural "progress" in its limited understanding of history and human life.

If such reclamation is to take place on a large scale, it is imperative to abandon any essentialist idea that we can "go back" to the agricultures destroyed by colonization or that if everyone farmed "like the Indians" all would be made right. Nostalgia is an impediment to rethinking our relationship to agriculture and each other. Likewise, to take the categories of agriculture as established by the colonists—nature, culture, tools, crops—and attempt to engineer them differently is to miss an important chance to reclaim the power to make and use the knowledge of food production and social interaction. As people seek alternatives to colonial agriculture and its industrial produce, there is a tendency, for example, to be satisfied with technological solutions. Organic agriculture that does not address the issues of land tenure and labor has not claimed very much for itself. Even a reconstitution of nineteenth-century agricultural techniques organized around a community's local needs and commitments does not gain much if it fails to address the patriarchal arrangement of nineteenth-century families.

Many of us are at an unquestionable disadvantage in attempting to remember any relationship among ourselves or between us, our food, our tools, and the land that has not been riven by the structures of colonization. Our very understanding of history—how we got to be who we are—has been marked by the same thing. The more privilege we have as heirs to colonial conquest, the less likely we are to know the first thing about food production, let alone egalitarian societies.

This does not diminish our responsibility to think about such things for ourselves. There is no archive where we can look up the best way to address problems of land concentration, soil erosion, rural poverty, groundwater poisoning, pesticide residues, global warming. Although each of these problems has been expressed in scientific terms, they are social as well as economic, agricultural, or technological. If we leave their solution to the "experts," whose sciences continually refuse to acknowledge their epistemological limitations and formally exclude points of view that draw attention to the made-up-ness of science, all of the complicated issues of social and environmental justice will be submerged in prescriptions for green industry, more wilderness, a bigger environmental regulatory apparatus. Perhaps these things are necessary among people whose social responsibility and working knowledge of the material world are so atrophied that they cannot imagine social, agricultural, technological, or environmental change taking place without some strong hand guiding their own. We can hope that this is not the case.

Notes

Introduction | Abduction

1. This etymological discussion is based on entries in the *Oxford English Dictionary*, hereafter referred to as the *OED*.

2. Jacques Derrida discusses the significance of the idea of the "proper" language of truth and ordinary speech and its relationship to "metaphoric" language in his essay, "White Mythology: Metaphor in the Text of Philosophy," in *Margins of Philosophy*, pp. 207–71. All language, he concludes in his explication of Aristotle's *Poetics*, is ultimately metaphoric, and all "ordinary" language and "proper" meanings of words therefore are less timeless or stable than they may appear. Indeed, our ideas about truth and the ordinary themselves are based on a deliberate mystification and devaluation of the "unreliable" linguistic means by which they were derived. The word Derrida chooses to designate the process of naming a concept (like truth or the proper) is "catachresis," a rhetorical term for creating a forced or inappropriate (and therefore not really "truthful") metaphor. Catachresis is a debased form of metaphor (or naming) in Aristotle's *Poetics*, but for Derrida, catachresis becomes the means by which all names as metaphors (including the ones ultimately understood to be beautiful or true or "effective") originate.

3. See Merchant, *Death of Nature*, esp. chap. 2.

4. Worster, "Transformations of the Earth" and "History as Natural History," both in *Wealth of Nature*, pp. 45–63, 30–44.

5. See, for example, Worster's discussion of the Frankfurt School in *Rivers of Empire*, pp. 53–55. Worster cites Karl Wittfogel often; he was one of the school's first members (although he did not remain associated with it) and a scholar of Chinese civilization and its systems of irrigation.

6. Ibid., p. 53.

7. Worster, "Transformations of the Earth," in *Wealth of Nature*, pp. 48–50.

8. Worster, "Seeing Beyond Culture," p. 1144. Worster was replying to several historians' responses to his article "Transformations of the Earth: Toward an Agroecological Perspective in History" (reprinted as "Transformations of the Earth" in *Wealth of Nature*), all of which appeared together as "A Roundtable: Environmental History" in *Journal of American History*. Worster's article was followed by Crosby, "An Enthusiastic Second"; Richard White, "Environmental History, Ecology, and Meaning"; Merchant, "Gender and Environmental History"; Cronon, "Modes of Prophecy and Production"; and Pyne, "Firestick History."

9. Worster, "New West, True West," in *Under Western Skies*, p. 24. Webb quotes Powell in an epigraph at the beginning of his book, *The Great Plains*, catching both writers in an unflattering pose: "The physical conditions which exist in that land, and which inexorably control the operations of men, are such that the industries of the West are necessarily unlike those of the East, and their institutions must be adapted to their industrial wants. It is thus that a new phase of Aryan civilization is being developed in the western half of America" (p. 2).

10. Worster, "New West, True West," p. 22.

11. Smith, *Virgin Land*, p. xi.

12. Strong, *Our Country*, p. 22. Strong addressed a list of "perils" that continue to preoccupy conservative discourse: immigration, socialism, "intemperance," and the city. Strong was a devout adherent of the theory of Anglo-Saxon racial superiority and believed the race would take over the world (ibid., pp. 174–75). He held a predictably negative view of immigrants, especially Catholics, but also numbered Mormons among the homegrown perils.

13. Smith, *Virgin Land*, p. ix.

14. Ibid., pp. vii, xi. Smith reiterated this point after having been taken to task by Barry Marks for forgetting his own injunction against measuring "myths" against "empirical reality" and finding them false. See Marks, "Concept of Myth in *Virgin Land*."

15. Kolodny, *Land Before Her*, p. xiii.

16. Ibid., p. ix.

17. Kuklick, "Myth and Symbol in American Studies." Actually, Kuklick appears to have taken the name of the school from a sentence in Smith's original preface (*Virgin Land*, p. xi): "The terms 'myth' and 'symbol' occur so often in the following pages that the reader deserves some warning about them." It was precisely Smith's theory and method (discussed in his preface) that identified the school, not the subject of the American West highlighted in Smith's title.

18. For historical accounts of American studies as a discipline, see Denning, "'Special American Conditions'"; Jehlen and Bercovitch, *Ideology and Classic American Literature*; Tate, *Search for a Method in American Studies*; and Wise, "'Paradigm Dramas.'" Russell Reising discusses many prominent American studies scholars in *Unusable Past*. Gerald Graff presents a history of literary studies, including the conventions and significance of New Criticism and some material on American studies scholars' contributions to American literary studies, in *Professing Literature*. For an example of the scientificity of New Criticism, see Frye, *Anatomy of Criticism*. Early American studies work on American literature besides Smith's *Virgin Land* includes Lewis, *American Adam*; Marx, *Machine in the Garden*; Matthiessen, *American Renaissance*; and Miller, *Errand into the Wilderness*.

19. Cronon, "Revisiting the Vanishing Frontier," p. 159. Ronald Carpenter's *Eloquence of Frederick Jackson Turner* exemplifies the interest in Turner's rhetoric that Cronon identifies. Turner's first academic job was teaching oratory.

20. Slotkin, *Fatal Environment*, pp. 40–41; Drinnon, *Facing West*, p. 461; Kolodny, *Lay of the Land*, pp. 136–37, and *Land Before Her*, pp. 3–5.

21. Smith describes his theoretical perspective in "Symbol and Idea in *Virgin Land*," in Jehlen and Bercovitch, *Ideology and Classic American Literature*, pp. 21–35, citing Jameson on p. 22. Marx cites Geertz in his account of the "myth and symbol school" in *Pilot and the Passenger*, p. x.

22. Smith, *Virgin Land*, p. viii.

23. Vaihinger, *Philosophy of "As If"*; Bergson, *Matter and Memory*; Lévy-Bruhl, *How Natives Think*. Smith's use of Lévy-Bruhl is interesting because he apparently adapted the latter's ideas about "primitive" people's conflation of themselves with their "totems" to a study of nineteenth-century America—an admission that people of the anthropologists' own societies can be anthropological objects.

24. Smith, *Virgin Land*, pp. vii–viii.

25. Merchant, *Death of Nature*, pp. xxi–xxii.

26. Ibid., p. xviii.

27. Merchant, *Ecological Revolutions*, p. 6.
28. Bookchin, *Ecology of Freedom*, p. 243.
29. Hallyn, *Poetic Structure of the World*, pp. 7–31.
30. Ibid., p. 12.
31. Ibid., p. 25.
32. Ibid., pp. 7–8.
33. Ibid., pp. 29–30.
34. Ibid., p. 31 (emphasis in original).
35. Ibid., p. 30.
36. Foucault, *Order of Things*, pp. xiii–xiv.
37. Hayden White, *Metahistory*, p. ix.
38. Nietzsche, "On Truth and Lies in a Nonmoral Sense," in *Philosophy and Truth*, p. 84.
39. Wright, "Political Institutions and the Frontier," p. 41.

Chapter 1 | Trees

1. Harrison, *Forests*, p. 1.
2. Ibid., p. 6.
3. Ibid., p. 51.
4. Ibid., p. 53.
5. Ibid., p. 55.
6. Gilles Deleuze and Félix Guattari use the idea of deterritorialization frequently and in many contexts throughout *Thousand Plateaus*. The structure of the idea is relatively simple, which is how it is easily transposed to different epistemological, political, or psychoanalytical problems. In order to gain control over something, you need to erase the codes that already control or direct the use or occupation of that thing and "recode," or reterritorialize, it. This is not in itself an authoritarian gesture as long as your recoding is not enforced as a permanent improvement. State colonization is at its core coercive, decoding and recoding its territory with the intention of resisting all efforts at decoding directed against it. Deterritorialization can challenge the hegemony of a system of laws or demands, but it is not by itself "liberatory."
7. For the state as "an instrument of organized violence," see Bookchin, *Ecology of Freedom*, p. 123. For Bookchin's discussion of authoritarian instruments of administration, see ibid., chaps. 4, 10. Bookchin understands "instruments of administration" to include an authoritarian epistemology as well as an authoritarian technology, enmeshing society in coercive, hierarchical relationships that are reinforced at every level of social, technological, ecological, and bureaucratic interaction.
8. Marsh, *Man and Nature*, p. 269.
9. On the development of silviculture in England, see Fernow, *History of Forestry*, pp. 365–67, and Greeley, *Forest Policy*, pp. 107–10. On the perception of unlimited timber supplies and its effect on delaying changes in lumber consumption and the development of silviculture in the United States, see Fernow, *History of Forestry*, pp. 466, 468, 470; Greeley, *Forest Policy*, pp. 145, 152–53; and Shirley Allen, *Introduction to American Forestry*, pp. 258–59.
10. Greeley, *Forest Policy*, p. 3.
11. Fernow, *History of Forestry*, pp. 1–2.
12. Shirley Allen, *Introduction to American Forestry*, p. 258.

13. Greeley, *Forest Policy*, p. 3.

14. Fernow, *History of Forestry*, p. 6.

15. Greeley, *Forest Policy*, p. 12.

16. Sparhawk, "History of Forestry in America," p. 702.

17. Shirley Allen (*Introduction to American Forestry*, p. 17) and Bernhard Fernow (*Economics of Forestry*, p. 371) both cite the fact that the population of the United States quadrupled between 1820 and 1870.

18. The public domain is more closely related to the concept of the medieval forest, as lands set aside for the sovereign, than it is to any notion of "the people's" land. The forest was generally set aside from lands not already claimed by ownership or usufruct. Neither the medieval forest nor the public domain was ever in practice a commons. See Harrison, *Forests*, and Fernow, *History of Forestry*, on the conventions of the medieval forest.

19. Shirley Allen, *Introduction to American Forestry*, pp. 264–65.

20. Simmons, "Yesterday and Today," p. 687. Simmons notes that Norse loggers sent cargoes of timber to Europe as early as 1000. Sparhawk, "History of Forestry in America," in the same yearbook, adds that timber exports from New England began before settlement, forming "the basis of a thriving trade with the West Indies and with Europe" in the early seventeenth century (p. 702).

21. Ise, *United States Forest Policy*, p. 26.

22. Greeley, *Forest Policy*, pp. 146–47.

23. See ibid., pp. 145–47; Simmons, "Yesterday and Today," pp. 687–89; and Fernow, *History of Forestry*, pp. 466–72.

24. Deleuze and Guattari, *Thousand Plateaus*, p. 385.

25. Ibid., p. 386.

26. Fernow, *Economics of Forestry*, p. 100.

27. Baker, "Silviculture," p. 67.

28. Ibid.

29. Fernow, *Economics of Forestry*, p. 101.

30. Baker, "Silviculture," p. 67.

31. Ibid., pp. 69–70.

32. Quoted in Ise, *United States Forest Policy*, pp. 16–17.

33. Ibid., p. 17.

34. Ibid., p. 20; Robbins, *American Forestry*, p. 5.

35. Watts, "U.S. Forest Service," p. 165.

36. Winters, "First Half Century," p. 7.

37. Ise, *United States Forest Policy*, pp. 114–17.

38. Pinchot, *Breaking New Ground*, p. 140.

39. Greeley, *Forests and Men*, p. 103.

40. Ibid.

41. Clepper, *Professional Forestry in the United States*, pp. 274–75.

42. Ibid., pp. 276–77.

43. Ibid., p. 273.

44. This theory has long been controversial, but I find the antistate quality of broadcast fire sufficient evidence of its use by societies "against the state," in Pierre Clastres's term. Stephen Pyne's account of American aboriginal fire practices, along with an account of the controversy regarding intentionally set fires, is in *Fire in America*, chap. 2.

45. Pyne describes the conflict between "old fashioned" land-clearing fire practices and the developing trend toward fire control throughout the nineteenth century in *Fire in America*, chap. 3.

46. Ibid., p. 199.

47. Ibid., p. 201.

48. Ibid., p. 202.

49. Ibid., p. 204.

50. Holbrook, *Burning an Empire*, p. 11.

51. Ibid., pp. 66, 15.

52. Firsthand accounts, quoted in Pyne, *Fire in America*, p. 206.

53. Ise, *United States Forest Policy*, p. 69.

54. Secretary of the Interior Ethan Hitchcock, quoted in Pyne, *Fire in America*, p. 230.

55. Clepper, *Professional Forestry in the United States*, p. 271.

56. Except where noted, this account of the 1910 fires is based on Cohen and Miller, *The Big Burn*. There are many other accounts of the fires, including Pyne, *Fire in America*; Holbrook, *Burning an Empire*, chap. 11; and Greeley, *Forests and Men*, chap. 1.

57. Holbrook, *Burning an Empire*, p. 128.

58. Timothy Cochrane, "Trial by Fire," p. 17.

59. Ibid.

60. MacLean, *Young Men and Fire*. The connection between national security and forest security, with respect to fire, is clear on the memorial plaque honoring those who died in the Mann Gulch fire: "IN MEMORY OF the 13 heroic young men who lost their lives in service of their country fighting the Mann Gulch forest fire 1 mile down the river on August 5, 1949" (ibid., p. 16).

61. Holbrook, *Burning an Empire*, p. 131.

62. Ibid., p. 123.

63. Greeley, *Forests and Men*, p. 23.

64. Pyne, *Fire in America*, p. 244.

65. Ibid., p. 255.

66. Greeley, *Forests and Men*, pp. 87–88; Winters, "First Half Century," p. 14.

67. Greeley, *Forests and Men*, pp. 87–89.

68. Cowan, *The Enemy Is Fire!*, p. 44.

69. Greeley, *Forest Policy*, p. 6.

70. Kauffman, "Spruce Goes Back to War," p. 364.

71. Graves, quoted in William G. Robbins, *American Forestry*, p. 113.

72. Ibid., p. 114.

73. Ibid.

74. Ise, *United States Forest Policy*, pp. 313–14.

75. Greeley, *Forests and Men*, p. 94.

76. William G. Robbins, *Hard Times in Paradise*, pp. 50–51.

77. Greeley, *Forests and Men*, pp. 93–94; Kauffman, "Spruce Goes Back to War," p. 364.

78. Cohen, *Tree Army*, pp. 90–91.

79. Roosevelt's message of 21 March 1933, quoted in ibid., p. 6.

80. Greeley, *Forests and Men*, pp. 140, 143.

81. Ibid., p. 140.

82. Civilian Conservation Corps et al., *CCC at Work*, pp. 30, 45.

83. Fechner, quoted in Salmond, *Civilian Conservation Corps*, p. 116.

84. McEntee, *Now They Are Men*, pp. 36, 64.

85. Salmond, *Civilian Conservation Corps*, pp. 118–19.

86. Hay, "For CCC Military Training," p. 242.

87. McEntee, *Now They Are Men*, pp. 48–49.

88. Cohen, *Tree Army*, p. vi.

89. Holbrook, "Forest Goes to War," p. 55.

90. "National Defense Lays Heavy Lumber Demand upon Industry," p. 475.

91. "Forestry in Congress," p. 520.

92. Donald Coleman, "Wood in Modern Warfare," pp. 8–10.

93. Ibid., p. 8.

94. Nelson Brown, "War and Wood in Germany," p. 587.

95. Donald Coleman, "Wood in Modern Warfare," p. 7; Nelson Brown, "War and Wood in Germany," pp. 588–89; Butler, "Forests and National Defense," p. 27.

96. Donald Coleman, "Wood in Modern Warfare," p. 8.

97. Butler, "Editor's Log" (June 1940), p. 245.

98. Kauffman, "Spruce Goes Back to War," p. 363.

99. Greeley, *Forests and Men*, p. 153.

100. Holbrook, "Forest Goes to War," p. 59.

101. Ibid.

102. "Conservation War Front," p. 109.

103. "First Spruce Raft Arrives from Alaska," p. 132; Greeley, *Forests and Men*, p. 153.

104. Holbrook, "Forest Goes to War," p. 58.

105. "Conservation War Front," p. 109.

106. Kauffman, "Spruce Goes Back to War," p. 364.

107. "Regulation Urgent, Says Forest Service," p. 134.

108. Butler, "Lumber on the Carpet," p. 121; "National Defense and Public Regulations," p. 126.

109. Granger, "National Forests at War," p. 112.

110. "First Spruce Raft Arrives from Alaska," p. 132.

111. Butler, "War and the Parks," p. 121.

112. "Need for Cut in Lumber Uses Weighed by WPB," p. 248.

113. See Nelson Brown, "War and Wood in Germany," p. 587; Butler, "Editor's Log" (March 1941), p. 101. Donald Coleman in "Wood in Modern Warfare" noted that "the prime motive of Hitler's early seizure of Poland is reported to lie in the abundance of wood offered by the vast forest resources of that country" (p. 7).

114. Butler, "Forests in the National Defense," p. 343.

115. Butler, "Editor's Log" (May 1941), p. 213.

116. James Montgomery Flagg poster, reproduced in *American Forests* 47 (May 1941): 240.

117. Butler, "Fire and Defense," p. 281.

118. Pacific Marine Supply Company advertisement, on back cover of *American Forests* 48 (July 1942).

119. John Clark Hunt, "If War Comes to the Forest," pp. 434, 407, 408.

120. Pyne, *Fire in America*, p. 287.

121. "Incendiary Fires Sweep California," p. 535.

122. Ovid Butler, Editorial, *American Forests* 48 (May 1942): 407. Pyne says this was

"largely self-imposed censorship" (*Fire in America*, p. 396), but *American Forests* editor Ovid Butler doesn't make it that clear. Considering that American foresters were working for industry and the federal government, as well as contributing to *American Forests*, any censorship might have been self-censorship.

123. Butler, "Editor's Log" (September 1941), p. 405.

124. Zahn, "San Diego Fires."

125. Pyne, *Fire in America*, p. 396.

126. Granger, "National Forests at War," p. 112; Woods, "Enemy Fire!" p. 232.

127. Pyne, *Fire in America*, p. 396.

128. Ibid., pp. 176–77.

129. Ibid., p. 174. Pyne discusses the ongoing relationship between western fire prevention, fire fighting, and war, so well coordinated during World War II, in *Fire in America*, chap. 7.

130. Holbrook, "Forest Goes to War," pp. 55–57.

131. Buckeye bulldozer advertisement, in *American Forests* 50 (July 1944): 325.

132. Winters, "First Half Century," p. 27.

133. Greeley, *Forests and Men*, p. 235.

134. Green, "Real Interest of the People," p. 754. Green was then president of the AFL.

135. Murray, "Labor Looks at Trees and Conservation," p. 757. Murray was then president of the CIO.

136. "On the Forest Fire Front," p. 247.

137. Holbrook, "Forty Men and a Fire," p. 251.

138. Ibid., p. 252.

139. Kerr, "Sky-Fighters of the Forest," p. 431.

140. John Clark Hunt, "Fire-Fighting after the War," p. 470.

141. George M. Hunt, "Forest Products Laboratory," p. 650.

142. Cline, "Future Requirements for Timber," p. 739.

143. Greeley, *Forests and Men*, pp. 156–57.

144. "Haiti to Plant Rubber for U.S.," p. 446.

145. Kauffman, "Guayule," p. 72; Cox, "Wild Rubber," p. 170.

146. Record and Hess, *Timbers of the New World*, pp. v–vi.

147. Cox, "Wild Rubber," p. 170.

148. Record and Hess, *Timbers of the New World*, pp. v–vi.

149. Ibid., p. xv.

150. Henius, "Oriente," p. 391.

151. Cox, "Wild Rubber," pp. 188–89.

152. Hays, *Conservation and the Gospel of Efficiency*, pp. 261–76.

Chapter 2 | Plows

1. Charles Little, introduction to Edward Faulkner's *Plowman's Folly*, p. xii.

2. Lynn White describes the European ard, which he calls a "scratch plough," as "essentially an enlarged digging stick dragged by a pair of oxen" (*Medieval Technology*, p. 41).

3. Mies, *Patriarchy and Accumulation*, p. 55.

4. Ibid., p. 57.

5. Martin Jones, "Regional Patterns in Crop Production," pp. 121–22.

6. Ibid., p. 123.

7. Solbrig and Solbrig, *So Shall You Reap*, pp. 44–45.

8. See Wilson, *Buffalo Bird Woman's Garden*, and Fussell, *Story of Corn*, pp. 99–113.

9. In addition to specific references noted in the text, my account of plow agriculture in medieval and Renaissance Europe is based on several sources: Merchant, *Death of Nature*, chap. 2; Bloch, *Feudal Society*, vol. 1, chap. 4; Braudel, *Capitalism and Material Life*, chaps. 2, 6; Lynn White, *Medieval Technology*, chaps. 2, 3; Bookchin, *Ecology of Freedom*, chaps. 3, 10; and Mies, *Patriarchy and Accumulation*, chap. 2.

10. Lynn White, *Medieval Technology*, p. 40.

11. Ibid., p. 41.

12. See *OED* entry for "lord."

13. Obviously this was not a uniform process. In places where warrior lords had never gained a foothold, rural people themselves controlled the practices of landownership and agricultural intensification, as they did in the Netherlands. As Carolyn Merchant notes, Dutch farmers had owned their land for centuries, and when they intensified their agriculture, they did so without sacrificing the sustainability of their practices or dispossessing people. (What the Dutch East India Company did at the South African Cape was, however, another story.) England presented a very different picture of rural immiserization. Germany had both pockets of noble control and pockets of relatively free farmers (Merchant, *Death of Nature*, pp. 51–56).

14. Braudel, *Capitalism and Material Life*, p. 89.

15. Ibid., p. 78.

16. Ibid., p. 73.

17. Sahlins, *Stone Age Economics*, p. 36.

18. Grieg, "Plant Resources," pp. 112–13.

19. Ibid., p. 114.

20. Tull, *Horse Hoeing Husbandry*, p. 7. Regarding vineyards, see ibid., pp. 66–68. Tull's method was based on the theory that soil is infinitely divisible and that if you cultivate the soil assiduously enough, you will always expose more "food" for the crops' "lacteal mouths," even without manure. Manure only serves to dissolve the soil, it does not furnish nutrients to the plants. Regarding plant nutrition, see ibid., pp. 9–19, 31–53.

21. For accounts of the development of American plow technology, see Hurt, *American Farm Tools*; Rogin, *Introduction of Farm Machinery*; Ardrey, *American Agricultural Implements*; and Holbrook, *Machines of Plenty*.

22. The South, with its "peculiar institution," did not participate in developments in plow technology. Southern agriculture, although thoroughly entrenched in the large-scale raising of cash crops for domestic and foreign markets, relied on coerced and captive labor rather than the labor of individual owners or tenants. Its clearings were maintained by hoes and by what Rogin calls a "shovel plow"—"something like a paring spade, that I do not think worth describing." The hoe was ubiquitous. Moldboard plows were used where there were few laborers and heavy soil. Shovel plows were much lighter implements, and interestingly the one image Rogin provides of this tool is a description of it in the hands of a girl: "I have seen a negro girl of fourteen or fifteen years old, mount her mule and take her plow on her shoulder or on before her, to ride to the field." Of course, the use of the hoe and the fieldwork of women in this context in no way constitute a "hoe agriculture" or women's

autonomous food production. On the contrary, plantation clearing and cultivation involved labor-intensive methods that, in the absence of captive laborers, were replaced by faster and more effective equipment for the profitability of individual owners. See Rogin, *Introduction of Farm Machinery*, pp. 53–54.

23. For historical accounts of how and why different groups of people took up western agriculture, see Katz, *The Black West*; Hurt, *Indian Agriculture in America*; Limerick, *Legacy of Conquest*; and Zinn, *People's History of the United States.*

24. See Hurt, *Indian Agriculture in America*, chaps. 1–5, and Wilson, *Buffalo Bird Woman's Garden*, pp. 15, 24. Buffalo Bird Woman reported that her family's largest field was about 540 feet by 270 feet, or a little more than 3 acres. In England in 1300, 5 acres or less was a small parcel of land for a peasant family; peasants might farm as much as 1,200 acres. In either case, the "arable" was used for field crops, not food, and was supplemented by smaller gardens closer to home (see Dyer, "Documentary Evidence," pp. 21–22). Before the Iron Age decline in food-plant production and increase in cereal-grain production, it is unlikely that 5 acres would have been insufficient to supply a small group of people.

25. The hope of African Americans in the South for "forty acres and a mule"—small farms on former plantations where they could raise both their own food and small amounts of commercial crops—was crushed in favor of their reinscription into the industrial agricultural economy as wage earners and tenants. We can only speculate how a European commitment to "forty acres"–style settlement would have affected European and American history.

26. See Limerick, *Legacy of Conquest*, chap. 2.

27. This account of bonanza farming is based on Hammer, "Bonanza Farming."

28. *Implement Blue Book*, pp. 155–81.

29. Webb, *The Great Plains*, p. 390.

30. Ibid., p. 391.

31. For the effects of wheat trading on the regional economy, see Cronon, *Nature's Metropolis*, chap. 3; for the development of Minneapolis as a milling capital, see Dan Morgan, *Merchants of Grain*, chap. 3. See also Blegen, *Building Minnesota*.

32. Marcus provides a history of the implementation of the Hatch Act in *Agricultural Science and the Quest for Legitimacy*. Willard Cochrane includes a history of the USDA's development of experiment stations and outreach in *Development of American Agriculture*, pp. 104–7, 243–57.

33. Newton, *Report of the Commissioner of Agriculture, 1862*, pp. 3–12.

34. Hurt, *Indian Agriculture in America*, pp. 32–33.

35. See ibid., chaps. 7, 8. "Agriculturalization," as I use it here, refers, of course, to the desired adoption by Native Americans of European tools, crops, methods, and social relations, *not* food production.

36. Indeed, as M. Annette Jaimes points out, "upwards of 60% of the subsistence of most Native American societies came directly from agriculture, with hunting and gathering providing a decidedly supplemental source of nutrients. . . . This highly developed agricultural base was greatly enhanced by extensive trade networks and food-storage techniques that afforded precontact American Indians what was (and might still be, if reconstituted) by far the most diversified and balanced diet on earth" ("Re-Visioning Native America," p. 10).

37. Holder, *Hoe and the Horse on the Plains*, pp. 58–59, 71, 77–81.

38. Hurt, *Indian Agriculture in America*, pp. 121–22.

39. Ibid., p. 46.

40. Castetter and Bell, *Pima and Papago Indian Agriculture*, pp. 152–53.

41. Hurt, *Indian Agriculture in America*, pp. 49–50.

42. Ibid., p. 125.

43. Noriega, "American Indian Education in the United States," pp. 378–85.

44. Hurt, *Indian Agriculture in America*, p. 127.

45. See Willard Cochrane, *Development of American Agriculture*, p. 93 (on agricultural prices), and pp. 107–8 (on mechanization).

46. See Webb, *The Great Plains*, pp. 152–60, and Smith, *Virgin Land*, chap. 16. See also Worster, *Rivers of Empire*, pp. 65–67.

47. Franklin, quoted in Smith, *Virgin Land*, p. 125.

48. Fite, *Farmers' Frontier*, p. 114.

49. Hilgard and Osterhout, *Agriculture for Schools of the Pacific Slope*, p. v.

50. Smith, *Virgin Land*, p. 178.

51. Ibid., p. 179.

52. Wilber, *Great Valleys and Prairies of Nebraska*, quoted in Smith, *Virgin Land*, p. 182.

53. James Wilson, Report of the Commissioner of Agriculture, in U.S. Department of Agriculture, *Annual Reports . . . 1905*, p. xxxviii. See also Hargreaves, *Dry Farming*, pp. 290–309, for a detailed and extremely well documented account of botanical investigations on behalf of western dry farming.

54. Widtsoe, *Dry-Farming*, pp. 234, 237. Spring wheat is sown in the spring; winter wheat is sown in the fall.

55. Ibid., p. 238. These Mennonite settlers appear many times in stories of northern Plains agricultural development. See also Hargreaves, *Dry Farming*, pp. 84, 296n, 451.

56. Widtsoe, *Dry-Farming*, p. 240.

57. Hargreaves, *Dry Farming*, pp. 83–84.

58. Ibid., pp. 85–94.

59. Atkinson, "Dry Farming Investigations in Montana," p. 65.

60. Linfield, "Dryland Farming in Montana," p. 12.

61. Widtsoe, *Dry-Farming*, pp. 301–20.

62. Hargreaves, *Dry Farming*, pp. 90–91.

63. See Widtsoe, *Dry-Farming*, p. 362.

64. Ibid., pp. 1–5.

65. Ibid., pp. 380–81.

66. See Willard Cochrane, *Development of American Agriculture*, pp. 85–89.

67. Hurt, *The Dust Bowl*, p. 27.

68. Widtsoe, *Dry-Farming*, pp. 301–2.

69. U.S. Department of Agriculture, *Annual Report of the Secretary of Agriculture, 1925*, pp. 54–55.

70. Willard Cochrane, *Development of American Agriculture*, p. 111.

71. See Malin, *History and Ecology*, pp. 45–60.

72. Worster, *Dust Bowl*, and *Nature's Economy*, chap. 12.

73. Hurt, *The Dust Bowl*, p. 30.

74. Willard Cochrane, *Development of American Agriculture*, pp. 124–26.

75. Hurt, *The Dust Bowl*, p. 154.

76. Willard Cochrane, *Development of American Agriculture*, p. 124.

77. Douglas Hurt describes the development of irrigation technology from prehistory to the present in *Agricultural Technology in the Twentieth Century*, chap. 4. Donald Worster describes the development of the West as a hydraulic society, using the work of Karl Wittfogel to inform his understanding of these societies, in *Rivers of Empire*. He emphasizes aridity as a factor determining the possibilities of social organization and water use in the West. Donald Pisani presents a history of western water law and policy that, in explicit contrast to Worster's, assumes that the West was "defined as much by American values, culture and institutions, as by climate and geography," and that "local economic conditions mattered far more in the evolution of western water policy than aridity *per se*" (*To Reclaim a Divided West*, p. xvi). Pisani examines the history of irrigation in California in *From the Family Farm to Agribusiness*.

78. Worster, *Rivers of Empire*, p. 51.

79. Corbett, *Garden Farming*, p. v.

80. Schilletter and Richey, *Textbook of General Horticulture*, p. 45.

81. Ibid., pp. 9, 36–38, 42–59; Henry Albert Jones and Rosa, *Truck Crop Plants*, p. xi.

82. Schilletter and Richey, *Textbook of General Horticulture*, pp. 42–60.

83. Ibid., pp. 40–41.

84. Pisani, *From the Family Farm to Agribusiness*, p. 442. In calculating the number of rural people this land could potentially support, Pisani uses 40 acres as the size of a single small farm. This is the size of a small farm in the context of commercial agriculture, certainly, but not necessarily in the context of food production.

85. See Gonzales, *Mexican and Mexican American Farm Workers*, pp. 1–20.

86. See Gamboa, *Mexican Labor and World War II*, p. 65, regarding the "stockpiling" of braceros in the 1940s. Growers' associations contracted workers and refused to release them to work elsewhere, even when the associations had no work for them to do.

87. Ibid., p. 90.

88. Gonzales, *Mexican and Mexican American Farm Workers*, p. 5.

89. Pisani, *From the Family Farm to Agribusiness*, pp. 440–43. Pisani cites a 1916 California commission on land colonization, from which these comments about ignorant labor and men and women of character and intelligence are taken.

90. Schilletter and Richey, *Textbook of General Horticulture*, p. 28.

91. See Tannahill, *Food in History*, pp. 310–12, regarding the history of canning and the processing of American fruits and vegetables. Cronon does not discuss the Chicago canneries in *Nature's Metropolis*.

92. Griffiths, Olcott, and Shaw, "Our Second Largest Food Group," pp. 213, 216.

93. Ruiz, *Cannery Women, Cannery Lives*, pp. 23–25.

94. Henry Albert Jones and Rosa, *Truck Crop Plants*, p. ix.

95. Schilletter and Richey, *Textbook of General Horticulture*, pp. vii, 1.

96. Kolodny, *Land Before Her*, p. xiii.

97. Ibid.

98. Norwood, *Made From This Earth*, p. 133.

99. Ibid., pp. 130–31, 142.

100. For an account of women's roles on American farms and the farm family economy in general, see Sachs, *Invisible Farmers*.

101. Greathouse, "Vegetable Garden," p. 3.

102. Vrooman, "Grain Farming in the Corn Belt," pp. 41–42.

103. Funk, "What the Farm Contributes Directly to the Farmer's Living," p. 7.

104. Vrooman, "Grain Farming in the Corn Belt," p. 41.

105. The Oliver Kelley Farm, established in Elk River, Minnesota, by the man who founded the Grange, is now a living-history farm maintained by the Minnesota Historical Society. Nineteenth-century tools, crops, costumes, and gender roles are featured at the farm. The kitchen garden is plowed and cultivated by men and weeded by women. The rows are long and straight and as far apart as six feet. The soil between the rows is loose and tends to be dry and sandy, unprotected from wind or water erosion. During the several seasons I visited the farm, the "garden" had none of the attributes of plenty or fecundity culturally associated with gardens; it was certainly not enclosed; and it was vast and dusty. Work there under the broiling sun in a humid river valley by a solitary housewife must surely have been drudgery. The museum women at least have each other for company and work sharing. Interestingly, among the twentieth-century men and women who work this nineteenth-century farm, there is an ongoing concern about the centrality of the men's fieldwork in the museum's educational program and the sense among some women that activities in the house and garden should be highlighted more. The museum has duplicated the marginalization of women's agricultural expertise by marginalizing the historical expertise of the women playing the women's roles.

106. Beal, "Farmer's Vegetable Garden," p. 7.

107. Keffer, "The Garden," p. 233.

108. Beattie, "Home Vegetable Garden," p. 8.

109. The earlier bulletin appeared in 1906. Beattie published the second bulletin with J. H. Beattie as "The Farm Garden" in 1931.

110. Wickson, "Irrigation in Field and Garden," pp. 33–36.

111. Tucker, *Kitchen Gardening in America*, pp. 72–91.

112. Clastres, *Society against the State*, p. 18.

113. Rosaldo, *Culture and Truth*, chap. 3.

114. All etymological discussion here is based on entries in the *OED*.

115. Crosby, *Ecological Imperialism*, p. 290.

116. Lynn White, *Medieval Technology*, p. 41.

117. Ardrey, *American Agricultural Implements*, p. 5.

118. Faulkner, *Plowman's Folly*, p. 5.

119. Morgan, "Greater America." Morgan was a former implement dealer, then a captain in the U.S. Army, writing from the Philippines.

120. Ardrey, *American Agricultural Implements*, p. 6.

121. Wheelhouse, *Digging Stick to Rotary Hoe*, pp. 13, 14.

122. Lynn White, *Medieval Technology*, p. 43.

123. Ives, "How the Russians Plow," p. 15.

124. Webb, *The Great Plains*, pp. 390–91.

125. Ardrey, *American Agricultural Implements*, pp. 8–9.

126. Ibid., p. 18.

127. Chase, "Farm Machinery," pp. 439–40 (emphasis in original).

128. Rogin, *Introduction of Farm Machinery*, p. 47.

129. Arthur W. Turner and Johnson, *Machines for the Farm, Ranch, and Plantation*, p. 33.

130. See Shiva, *Violence of the Green Revolution*.

Chapter 3 | Grass

1. Sampson, *Range Management*, p. 361.

2. Accounts of the stock boom of the late nineteenth century are predictably numerous. A classic version, at once a heroic story and part of a larger scholarly historical account of the Plains, is Webb, *The Great Plains*, chap. 6. Crosby discusses the cattle industry in the context of the "portmanteau biota" carried by Europeans into colonized regions that became "neo-Europes" in *Ecological Imperialism*, pp. 279–80. Cronon presents the story of the open range, with many bibliographical citations, in *Nature's Metropolis*, pp. 218–24. Cronon reports that "still the best account of the cattle drives is Andy Adams, *The Log of a Cowboy: A Narrative of the Old Trail Days* (1903; reprint, 1964)" (*Nature's Metropolis*, p. 438 [n. 45]).

3. Edwards, "Settlement of the Grasslands," p. 30.

4. Kraenzel, *Great Plains in Transition*, p. 84.

5. Frederick Jackson Turner, "Significance of the Frontier in American History," p. 6.

6. Billington, *Westward Movement in the United States*, pp. 9–10.

7. Fite, *Farmers' Frontier*, p. v.

8. Webb, *The Great Plains*, p. 239.

9. Johnson, *Farm Animals in the Making of America*, pp. 64, 77.

10. Osgood, *Day of the Cattleman*, p. 27.

11. Sandoz, *Cattlemen from the Rio Grande*, p. 35.

12. Kraenzel, *Great Plains in Transition*, p. 105.

13. James MacDonald, *Food from the Far West*, quoted in Osgood, *Day of the Cattleman*, pp. 26–27.

14. Way and Simmons, *Geography of Spain and Portugal*, p. 142.

15. On Andalusian cattle of all kinds, see ibid. and Wellman, *The Trampling Herd*, p. 18.

16. Wellman, *The Trampling Herd*, p. 21.

17. Shaw, *Study of Breeds in America*, p. 3.

18. L. F. Allen, "Shorthorn Breed of Cattle," pp. 417–22.

19. Sandoz, *Cattlemen from the Rio Grande*, p. 186. Tayo, the protagonist of Leslie Marmon Silko's novel *Ceremony*, understands the difference between Mexican cattle (closely akin to longhorns) and Anglo cattle in similar terms. After leaving the windmill where they had been unloaded, the Mexican cows he is looking after "would travel until they found more water. Herefords would not look for water. When a windmill broke down or a pool went dry, Tayo had seen them standing and waiting patiently for a truck or wagon loaded with water, or for riders to herd them to water. If nobody came and there was no snow or rain, then they died there, still waiting. But these Mexican cattle were different" (p. 79).

20. For an extensive account of the history of English cattle and their breeders in England and the United States, see Briggs and Briggs, *Modern Breeds of Livestock*, pp. 39–96.

21. Johnson, *Farm Animals in the Making of America*, p. 87.

22. Briggs and Briggs, *Modern Breeds of Livestock*, p. 66.

23. Shaw, *Study of Breeds in America*, pp. 23–30; Briggs and Briggs, *Modern Breeds of Livestock*, pp. 42, 79; Wellman, *The Trampling Herd*, pp. 288–89.

24. Osgood, *Day of the Cattleman*, p. 138n; Briggs and Briggs, *Modern Breeds of Livestock*, pp. 52, 80; Shaw, *Study of Breeds in America*, p. 42.

25. Wellman, *The Trampling Herd*, pp. 251–52, 289–90.

26. Ibid., p. 290.

27. Sandoz, *Cattlemen from the Rio Grande*, p. 486.

28. Way and Simmons, *Geography of Spain and Portugal*, pp. 139–40.

29. McAlister, *Spain and Portugal in the New World*, p. 216.

30. Ibid., p. 21. Other historians attribute the origins of the merino to the development of stockraising and wool production by North Africans in Spain (see Briggs and Briggs, *Modern Breeds of Livestock*, p. 427, and Wentworth, *America's Sheep Trails*, p. 10). The thirteenth-century expansion of wool production took place during the Spanish reconquest of the Peninsula after centuries of developing high-quality wool under the influence of North African breeders. Before that, Roman conquerors had developed Spanish sheep as well. The Greek historian and geographer Strabo, writing near the beginning of the first millennium A.D., believed that even then the finest wool in the world came from Iberia (ibid.). All of these waves of sheep breeding either benefited a colonial occupation or promoted the development of a very wealthy commercial class.

31. Wentworth, *America's Sheep Trails*, p. 11; Towne and Wentworth, *Shepherd's Empire*, p. 7.

32. McAlister, *Spain and Portugal in the New World*, pp. 219–20; Shaw, *Study of Breeds in America*, p. 175.

33. Shaw, *Study of Breeds in America*, p. 183; Briggs and Briggs, *Modern Breeds of Livestock*, pp. 427–28.

34. Wentworth, *America's Sheep Trails*, pp. 11–12.

35. Ibid., pp. 10–11.

36. Clarke, *Modern Sheep*, p. 12.

37. Shaw, *Study of Breeds in America*, p. 185.

38. See Towne and Wentworth, *Shepherd's Empire*, chaps. 2, 3.

39. Wentworth, *America's Sheep Trails*, p. 273.

40. Ibid., p. 135; Towne and Wentworth, *Shepherd's Empire*, p. 126.

41. Wentworth, *America's Sheep Trails*, p. 123.

42. Briggs and Briggs, *Modern Breeds of Livestock*, p. 441.

43. Wellman, *The Trampling Herd*, p. 329.

44. Ibid., pp. 329–30.

45. Osgood, *Day of the Cattleman*, p. 39.

46. Ibid., p. 189.

47. Sandoz, *Cattlemen from the Rio Grande*, pp. 461–62.

48. Osgood, *Day of the Cattleman*, pp. 229–30, 255. For accounts of sheep drives and cattle-sheep wars, see Wentworth, *America's Sheep Trails*, chaps. 13, 14; Towne and Wentworth, *Shepherd's Empire*, chaps. 7, 8; Wellman, *The Trampling Herd*, chap. 35; and Sandoz, *Cattlemen from the Rio Grande*, p. 253.

49. Potter, *Western Live-Stock Management*, pp. 54–58. See also Coffey, *Livestock Management*, pp. 74–75, and Widmer, *Practical Animal Husbandry*, p. 121.

50. Ensminger, *Sheep Husbandry*, pp. 113–14; Horlacher and Hammonds, *Sheep*, p. 67; Clarke, *Modern Sheep*, pp. 194–95. In 1886 Alexander Graham Bell began a fascinating experiment to increase the prolificacy of sheep by selecting ewes for multiple nipples (ordinary sheep have only two), reasoning that other mammals with multiple nipples regularly bear multiple offspring in a single birth. By 1923, his ewes had between five and six

nipples apiece, and multiple births had increased by 50 percent. The "extra" nipples did not give milk. See Wentworth, *America's Sheep Trails*, p. 553.

51. Nelson, "Sheep," pp. 175–81.

52. Wentworth, *America's Sheep Trails*, p. 546.

53. Charles Amsden, quoted in ibid., p. 548.

54. Ibid., pp. 548–49; Parman, *Navajos and the New Deal*, pp. 128–29. Wentworth cites Cecil Blunn, a sheep researcher who worked on the Navajo reservation in the 1930s and wrote an account of the history of Navajo sheep breeding before that time. See Blunn, "Improvement of the Navajo Sheep." Parman cites George M. Sidwell, Jack L. Ruttle, and Earl E. Ray, "Improvement of Navajo Sheep," Las Cruces, New Mexico, State University Agricultural Experiment Station Research Report #172, 1970, as a more recent and comprehensive history.

55. Parman, *Navajos and the New Deal*, p. 130.

56. Wentworth, *America's Sheep Trails*, p. 549.

57. Widmer, *Practical Animal Husbandry*, pp. 118–19.

58. Coffey, *Livestock Management*, p. 76.

59. Widmer, *Practical Animal Husbandry*, p. 120.

60. Coffey, *Livestock Management*, pp. 76–77.

61. This is most true, of course, for mutton-type sheep. Clarke put it this way, regarding Shropshire sheep in particular: "They carry a leg at each corner" (*Modern Sheep*, p. 18). In general, according to Horlacher and Hammonds, a mutton-type ewe should have "a straight back, wide loin, wide spring of ribs, deep body, and well-developed leg of mutton." A mutton-type ram should be "broad, deep, thick, straight on top, and close to the ground" (*Sheep*, pp. 67, 72). Ensminger uses similar terms to describe ideal mutton sheep (*Sheep Husbandry*, pp. 97–98). Sheep bred for fleece tend to be "more upstanding and angular, with considerably less width, depth, fullness and smoothness throughout," according to Ensminger, but the Rambouillet is of "much more acceptable mutton type than the extreme wool type of former years" (ibid., p. 99).

62. Horlacher and Hammonds, *Sheep*, pp. 66–68, 72–73, 92. See also Clarke, *Modern Sheep*, p. 180.

63. Clarke, *Modern Sheep*, p. 179.

64. Ibid., p. 18.

65. Briggs and Briggs, *Modern Breeds of Livestock*, p. 32.

66. Henry and Morrison, *Feeds and Feeding*, p. vii. Nine printings of this book were produced between 1898, when it first appeared, and 1910, when it was revised. I cite a 1915 edition.

67. Sampson, *Range and Pasture Management*, p. 3.

68. Armsby, *Nutrition of Farm Animals*, p. xv.

69. On the history of American eating habits and nutritional theory, see Harvey Levenstein, *Revolution at the Table* and *Paradox of Plenty*.

70. Senate report, quoted in Osgood, *Day of the Cattleman*, p. 216.

71. Norman Coleman, *Report of the Commissioner of Agriculture, 1887*, p. 8.

72. Osgood, *Day of the Cattleman*, p. 216; Sandoz, *Cattlemen from the Rio Grande*, p. 267.

73. Webb, *The Great Plains*, p. 237.

74. Osgood, *Day of the Cattleman*, pp. 94–96; Webb, *The Great Plains*, pp. 236–37.

75. Osgood, *Day of the Cattleman*, p. 228.

76. Ibid., p. 233.

77. Ibid., p. 236.

78. Spillman, "Grass and Forage Plant Investigations," p. 114.

79. Savage, Smith, and Costello, "Dry-Land Pastures on the Plains," p. 520.

80. Boardman, "Sheep Husbandry in the West," p. 288.

81. Piper, *Forage Plants*, p. 135.

82. Wooten and Barnes, "Billion Acres of Grasslands," p. 30.

83. Osgood, *Day of the Cattleman*, p. 236.

84. U.S. Department of Agriculture, *[Annual] Report, 1888*, pp. 31–32.

85. McKee, "Legumes of Many Uses," p. 715.

86. Spillman, "Grass and Forage Plant Investigations," pp. 111–13; Coville, "Bureau of Plant Industry Report," p. 305.

87. Keller and Hochmuth, "Cultivated Forage Crops," pp. 548–51. The region in question included all of Idaho, Nevada, Utah, and Arizona; eastern portions of California, Oregon, and Washington; and parts of western Montana, Wyoming, Colorado, New Mexico, and Texas.

88. Hoover et al., "Main Grasses for Farm and Home," p. 640.

89. Ibid., p. 642.

90. Savage, Smith, and Costello, "Dry-Land Pastures on the Plains," p. 507; Hoover et al., "Main Grasses for Farm and Home," p. 670.

91. Wheeler, *Forage and Pasture Crops*, pp. 245–46. See also Piper, *Forage Plants*, pp. 348–50, and Coburn, *Book of Alfalfa*, pp. 1–2, 4, 9. The word "alfalfa" has several Arabic variants, including "al-fasfasha." The British call alfalfa "lucerne," either from the valley of Lucerna in northwestern Italy or from *laouzerdo* or *lauserne*, French variants of the Catalonian *userdas*. *Userdas* is of uncertain origin. It designates the plant known as alfalfa but may be derived from words meaning light or flame. An Indo-European root appears in Old High German and Middle Irish words meaning meadow, grass, or fresh pasture and in Sanskrit and Persian words meaning springtime, suggesting a possible etymology for *userdas*. Wheeler claims that alfalfa spread from Italy to other European countries, including Spain. Piper notes that historical evidence supports the notion that Spain rather than Italy was the source of northern European alfalfa. The Spanish language has adopted many Arabic words for agricultural things, including *acequia* (irrigation canal) and *naranja* (orange), and it seems likely that "alfalfa" would have been included in the North African legacy as well. Coburn came to this conclusion in *The Book of Alfalfa*. Also, wheat growing was more important on the Roman latifundias in Spain than stock growing, which was one of the North Africans' specialties (see Way and Simmons, *Geography of Spain and Portugal*, p. 139). It seems more probable that a forage crop would be introduced by conquerors interested in promoting stockraising, explaining the survival of "alfalfa" rather than some version of "lucerne" in Spanish. See *OED*, entry for "lucerne"; Joan Coromines, *Diccionari Etimològic i Complementari de la Llengua Catalonia*, entry for "userda"; and Stuart E. Mann, *An Indo-European Comparative Dictionary*.

92. Piper, *Forage Plants*, p. 348.

93. Keller and Hochmuth, "Cultivated Forage Crops," p. 548.

94. See entries for these plants in U.S. Department of Agriculture, Forest Service, *Range Plant Handbook*; Hitchcock, *Manual of the Grasses*; Lauren Brown, *Grasslands*; Hoover et al., "Main Grasses for Farm and Home"; and McKee, "Legumes of Many Uses."

95. U.S. Department of Agriculture, Forest Service, *Range Plant Handbook*, p. 158.

96. Ibid., p. 70.

97. Mack, "Invasion of *Bromus tectorum*," p. 153.

98. U.S. Department of Agriculture, Forest Service, *Range Plant Handbook*, p. 542.

99. McKee, "Legumes of Many Uses," p. 722.

100. U.S. Department of Agriculture, Forest Service, Division of Range Research, "History of Western Range Research," p. 138.

101. U.S. Department of Agriculture, *Annual Reports . . . 1920*, p. 253.

102. See, for example, the recommendations made by Barnes, *Western Grazing Grounds and Forest Ranges*, p. 244. To counter this prevailing view, Sampson wrote in 1923 that overgrazing was *not* easy to judge only by the density of the plant cover or by the condition of the stock (*Range and Pasture Management*, pp. 104–5). Sampson's work, of course, attempted to define range condition more precisely.

103. Rowley, *U.S. Forest Service Grazing and Rangelands*, pp. 69–70.

104. Sampson, *Range and Pasture Management*, pp. 72–74. Sampson cited his own research to support his argument.

105. U.S. Department of Agriculture, Forest Service, Division of Range Research, "History of Western Range Research," p. 132; Alexander, "From Rule of Thumb to Scientific Range Management," pp. 409–28. The phrase "rule of thumb" as applied to prereconnaissance management appears to come from Clapp, "Major Range Problems." In this article, Clapp explains that range depletion was the result of the profit motive among stockraisers and poor "rule-of-thumb" management (p. 9).

106. Rowley, *U.S. Forest Service Grazing and Rangelands*, p. 101.

107. Ibid., p. 100.

108. Campbell, "Milestones in Range Management," p. 7.

109. Sampson, "Suggestions for Instruction in Range Management."

110. Clements, *Dynamics of Vegetation*, pp. 1–2, 9 (regarding definitions of succession and its analogue in human history); 99–118 (regarding plant indicators of overgrazing); 119–43 (regarding the nature and structure of the climax). Pyne wrote, "The concept of the Nebraskan Clements, in fact, bears an uncanny resemblance to the frontier hypothesis of Frederick Jackson Turner, according to which an area advances from wilderness to a pioneer stage to civilization" (*Fire in America*, p. 492). Worster discusses Clements's understanding of ecological/cultural succession in *Nature's Economy*: "Clements apparently never appreciated the considerable irony of this exclusion [of people from the ecological community]. . . . But the two processes of development [natural and cultural] were fated to meet, it seemed, in irreconcilable conflict. One would have to give way to another; it was not possible to have both a climax state of vegetation and a highly developed human culture on the same territory" (pp. 218–19).

111. Sampson, "Succession as a Factor in Range Management."

112. Sampson, *Range Management*, p. 5.

113. Ibid., p. 363.

114. Sampson, "Succession as a Factor in Range Management."

115. Talbot and Crafts, "Lag in Research and Extension," p. 186.

116. Ibid., p. 189.

117. Clapp, "Major Range Problems," fig. 15, p. 30; Bergoffen, "Questions and Answers," p. 33.

118. U.S. Department of Agriculture, Forest Service, Division of Range Research, "History of Western Range Research," p. 137.

119. Rowley, *U.S. Forest Service Grazing and Rangelands*, pp. 112–13.

120. U.S. Department of Agriculture, Forest Service, Division of Range Research, "History of Western Range Research," p. 143.

121. Clapp, "Major Range Problems," p. 30.

122. Ibid., p. 35.

123. Rowley, *U.S. Forest Service Grazing and Rangelands*, p. 157.

124. West, "USDA Forest Service Management of the National Grasslands," p. 87. Over 3 million acres of these repurchased lands became national grasslands in 1960.

125. Rowley describes the calculation this way: "The range survey provided critical information on the 'forage-acre-factor' for the writing of a range-management plan. A forage-acre was an ideal surface acre having 10/10 'palatable vegetation' density and a 100 percent palatability, the highest possible rating. In short, the forage-acre was a unit of measurement applied to the density and palatability of grasses, grasslike plants, and other herbaceous plants and available or grazable browse contained in each vegetative type. A palatability table provided by the regional office gave the palatability of individual plants and for all classes of stock. By multiplying the percentage of each individual by the palatability number the surveyors determined percentages of palatability of individuals, which, when added, gave the volume palatability of the type. This volume palatability multiplied by the density (estimated to hundredths, that is, 0.30, 0.35, and so on) yielded the aforementioned 'forage-acre-factor' more commonly known as FAF. For example, a sheep utilized 0.3 forage-acres per month, and a cow 0.8 forage-acres per month. On this basis the actual animal-unit-month (AUM) in any given type, unit, or forest could be figured, and when the length of season was determined, the number of stock which any area would support for the grazing season could be determined" (*U.S. Forest Service Grazing and Rangelands*, p. 164).

126. Ibid., p. 165.

127. See Voigt, *Public Grazing Lands*, pp. 278–322.

128. Rowley, *U.S. Forest Service Grazing and Rangelands*, p. 172.

129. Hurt, "National Grasslands," p. 258.

130. Pechanec, "Our Range Society," p. 1.

131. Worster, *Under Western Skies*, p. 49.

132. Ibid., pp. 28–29, 32.

133. Ibid., p. 93.

134. Webb, *The Great Plains*, pp. 10–44.

135. Ibid., pp. 32, 207, 395.

136. See Worster, *Nature's Economy*, p. 252. Although Webb does not use an explicitly ecological vocabulary, his ideas about the use of grasslands are generally consonant with those of range ecologists like Sampson. Worster provides an account of plant succession and climax communities in ibid., pp. 205–53.

137. Roessel's foreword, in Roessel and Johnson, *Navajo Livestock Reduction*, p. ix.

138. Eli Gorman's narrative, in ibid., pp. 26–27. *Diné* is the name the Navajo call themselves in their own language.

139. Martin Johnson's narrative, in ibid., p. 94.

140. Buck Austin's narrative, in ibid., p. 18.

141. Capiton Benally's narrative, in ibid., pp. 32–33.

142. Howard W. Gorman's narrative, in ibid., p. 44.

143. Roessel's foreword, in ibid., p. xi.

144. See Clapp, "Major Range Problems," pp. 29–31.

145. Hurt cites the acreage of Navajo rangeland that had been overgrazed in *Indian Agriculture in America*, p. 179. Lawrence C. Kelly notes that the Navajo reservation was the location of the Soil Erosion Service's first project in "Anthropology in the Soil Conservation Service." The Soil Erosion Service was established in 1933 and was replaced in 1935 by the Soil Conservation Service. For an overview of the Navajo range rehabilitation project as part of the Indian New Deal, see Parman, *Navajos and the New Deal.*

146. Hurt, *Indian Agriculture in America*, p. 180; Kelly, "Anthropology in the Soil Conservation Service," p. 139; Parman, *Navajos and the New Deal*, pp. 51–80.

147. Clapp, "Major Range Problems," p. 56.

148. Hurt, *Indian Agriculture in America*, pp. 176–77, 181–83.

149. See Worster, *Under Western Skies*, pp. 34–52. Worster discusses Törbel on pp. 39–40.

Chapter 4 | Weeds

1. Darlington, *American Weeds and Useful Plants*, p. xiii. The changes the editor of this edition made to Darlington's 1847 text involved updating the botanical arrangement and the names of plants (with the help of Asa Gray's work in 1859), as well as adding other plants not treated by Darlington (see ibid., p. vii).

2. Crosby, *Ecological Imperialism*, pp. 153, 154.

3. Ibid., p. 270.

4. Darlington, *American Weeds and Useful Plants*, pp. ix–x.

5. Ibid., p. xiii.

6. Ibid., pp. xiii–xiv.

7. Pammel, *Weeds of the Farm and Garden*, pp. 1–5, 87–103.

8. Georgia, *Manual of Weeds*, pp. 6–8.

9. Muenscher, *Weeds*, p. 26.

10. Ibid., pp. 53–54.

11. Wilfred W. Robbins, Crafts, and Raynor, *Weed Control*, p. 17.

12. Spencer, *All About Weeds*, p. 1.

13. Georgia, *Manual of Weeds*, p. ix.

14. "Noncommercial value" is not determined by whether a plant can be "used," like wheat; even ornamental plants have commercial value, whether when the nursery sells you a snowball bush for the corner of your yard or when the realtor appraises the beauty of your property in calculating the price of your house. A plant's noncommercial value might include the fact that it was your great-grandmother's, or that it should be placed over the front door on a specific night of the year, or that your people consider it holy and have always left it in the fields. Sometimes a plant's noncommercial value works to enhance the growth of more commercially valuable plants, such as, for example, when a child picks all the dandelions in a yard before they go to seed. These examples appear trivial only from a point of view that understands a plant's "usefulness" in the most narrow, commercial terms, dependent on successful monocultural agriculture for food or income.

15. Stein, *My Weeds*, p. 14.

16. Foy, Forney, and Cooley, "History of Weed Introductions," p. 69.

17. Ibid., p. 70.

18. Stein, *My Weeds*, p. 21. Tares are ryegrass, a species of *Lolium*.

19. Foy, Forney, and Cooley, "History of Weed Introductions," p. 70.

20. Stein, *My Weeds*, p. 22.

21. Foy, Forney, and Cooley, "History of Weed Introductions," p. 69.

22. Ridley, *Dispersal of Plants*, p. 630.

23. Nineteen of the twenty-nine most prevalent weeds in U.S. crops are species introduced in the last few centuries, according to USDA figures cited in Foy, Forney, and Cooley, "History of Weed Introductions," p. 68.

24. Crosby, *Ecological Imperialism*, pp. 21n, 332.

25. Ridley, *Dispersal of Plants*, pp. 630–31.

26. Foy, Forney, and Cooley, "History of Weed Introductions," p. 79.

27. See, for example, Crosby, *Ecological Imperialism*, p. 156; Haughton, *Green Immigrants*, pp. 282–85; Pamela Jones, *Just Weeds*, pp. 155–61; Coon, *Using Plants for Healing*, p. 158; and Darlington, *American Weeds and Useful Plants*, p. 219.

28. Jones, *Just Weeds*, p. 158.

29. Darlington, *American Weeds and Useful Plants*, p. 219. Regarding plantain as "dooryard plantain," see Muenscher, *Weeds*, p. 431.

30. Whitson, *Weeds of the West*, p. 405.

31. U.S. Department of Agriculture, Agricultural Research Service, *Common Weeds of the United States*, p. 347; Harrington, *Edible Native Plants of the Rocky Mountains*, pp. 84–86. Harrington gives "Indian wheat" as a common name for plantain.

32. Coon, *Using Plants for Healing*, p. 158; Pamela Jones, *Just Weeds*, p. 161. One of plantain's uses in Europe was as a wound dressing as well, although this use appears to have been abandoned among Euro-Americans.

33. Moerman, *Medicinal Plants of Native America*, 2:634.

34. Ibid., 1:350–52. The societies listed are Cherokee, Chippewa (separate from Ojibwa), Costanoan, Delaware (in Oklahoma), Delaware (in Ontario), Fox, Kawaiisu, Kwakiutl, Mahuna, Mohegan, Navaho-Ramah, Ojibwa, Paiute, Ponca, Potawatomi, Rappahannock, Shinnecock, Shoshone, and Shuswap.

35. Conversation with Rachel Buff about immigration policy, biology, and the ideology of the family brought these homologies with weed naturalization to mind.

36. Foy, Forney, and Cooley, "History of Weed Introductions," p. 76. Ridley mentions crop seeds first among all sources of weed introductions, noting that the "number of seeds of herbaceous plants carried about and planted with grain and other cultural plants is very large" (*Dispersal of Plants*, pp. 639–40).

37. Darlington, *American Weeds and Useful Plants*, p. 220.

38. Whitson, *Weeds of the West*, p. 109.

39. Dewey, "Canada Thistle" and "Migration of Weeds," p. 275. Dewey's version of the story appears in Wilfred W. Robbins, Crafts, and Raynor, *Weed Control*, p. 9. Dewey's account is quoted but not fully repeated in Pammel, *Weeds of the Farm and Garden*, p. 83.

40. Linnaeus, quoted in Latin and translated by Darlington in *American Weeds and Useful Plants*, p. 199.

41. See Muenscher, *Weeds*, p. 476.

42. Darlington, *American Weeds and Useful Plants*, pp. 197, 199.

43. Forcella and Harvey, *New and Exotic Weeds of Montana*, 2:34.

44. Wilfred W. Robbins, Crafts, and Raynor, *Weed Control*, p. 9.

45. See Pammel, *Weeds of the Farm and Garden*, pp. 28, 33, 35, 36.

46. Moerman only mentions Mohegan, Montagnais, and Ojibwa (Anishinabe) (*Medicinal Plants of Native America*, 1:122).

47. Muenscher, *Weeds*, p. 475.

48. Spencer, *All About Weeds*, p. 295.

49. Whitson, *Weeds of the West*, p. 109.

50. Spencer, *All About Weeds*, p. 293.

51. Moerman, *Medicinal Plants of Native America*, 1:122.

52. Wilfred W. Robbins, Crafts, and Raynor, *Weed Control*, p. 99.

53. Klingman, Ashton, and Noordhoff, *Weed Science*, pp. 205, 348, 352.

54. Muenscher, *Weeds*, p. 323. Poinsettia (*Euphorbia pulcherrima*) is a perhaps more familiar spurge, with characteristic milky sap.

55. *Leafy Spurge*, p. 7; Darlington, *American Weeds and Useful Plants*, p. 289.

56. Georgia, *Manual of Weeds*, p. 271.

57. Ibid.

58. Wilfred W. Robbins, Crafts, and Raynor, *Weed Control*, p. 466.

59. Georgia, *Manual of Weeds*, p. 271; Muenscher, *Weeds*, p. 324; Wilfred W. Robbins, Crafts, and Raynor, *Weed Control*, p. 467.

60. *Leafy Spurge*, p. 65.

61. Jim Krall, "Why I Didn't Get Rid of Leafy Spurge in Montana in 1952," in *Leafy Spurge*.

62. U.S. Department of Agriculture, Agricultural Research Service, *Common Weeds of the United States*, p. 248; *Leafy Spurge*, p. 2; Whitson, *Weeds of the West*, p. 317.

63. Whitson, *Weeds of the West*, p. 85; "Collection History of *Centaureas*." Regarding yellow starthistle, see Pammel, *Weeds of the Farm and Garden*, p. 250; regarding spotted knapweed, see Whitson, *Weeds of the West*, p. 89; regarding Russian knapweed, see Muenscher, *Weeds*, p. 469.

64. A. K. Watson and Renney, "Biology of Canadian Weeds," p. 693.

65. Forcella and Harvey, *New and Exotic Weeds of Montana*, maps, 2:29, 30. Spotted knapweed was first collected sometime between 1901 and 1910 in one county in western Montana. Russian knapweed was found in several counties in Washington and one county in northwestern Wyoming before 1920. Whitson, *Weeds of the West*, gives 1898 as the probable date of introduction of Russian thistle without providing further details (p. 93).

66. Muenscher, *Weeds*, p. 469.

67. Ibid., p. 471; Georgia, *Manual of Weeds*, p. 518. Regarding collection data of yellow starthistle, see Forcella and Harvey, *New and Exotic Weeds of Montana*, map, 2:31.

68. Whitson, *Weeds of the West*, p. 89; Fletcher and Renney, "Growth Inhibitor Found in *Centaurea*."

69. A. K. Watson and Renney, "Biology of Canadian Weeds," p. 687; U.S. Department of Agriculture, Agricultural Research Service, *Common Weeds of the United States*, pp. 384, 386.

70. Forcella and Harvey, *New and Exotic Weeds of Montana*, graphs, 2:102–3.

71. Muenscher, *Weeds*, p. 469.

72. Ibid., p. 468.

73. Wilfred W. Robbins, Crafts, and Raynor recommended a season or more of tillage to remove Russian knapweed in 1942 (*Weed Control*, p. 110).

74. Lacey et al., "Bounty Programs," p. 196.

75. Unless otherwise noted, this account of Russian thistle is based on James A. Young, "Tumbleweed."

76. Estimates of the date of Russian thistle's introduction vary. James A. Young, using the records of USDA botanist Lyster Dewey, writes that the plant was introduced "about 1877 on a farm in Bon Homme County, S.D." ("Tumbleweed," p. 84). Wilfred W. Robbins, Crafts, and Raynor wrote in 1942: "This weed was first introduced into the United States in 1873 or 1874 in flaxseed brought from Russia and sown near Scotland, Bonhomme County, South Dakota" (*Weed Control*, pp. 3–4). They do not cite Dewey regarding Russian thistle and give no other citation. In his 1897 essay, "Migration of Weeds," Dewey wrote that Russian thistle had been introduced less than twenty-five years before, or sometime after 1871 (p. 282).

77. Haughton, *Green Immigrants*, p. 399. Haughton also notes that *kali* is from the Arabic *alqaliy* (alkali), or the ashes of this plant. The Arabic name for the plant combined with the implication that it was burned to produce alkali suggest a history of its use among Arabic people independent of its presence in Russian flax fields.

78. Dewey, "Migration of Weeds," p. 270.

79. U.S. Department of Agriculture, Forest Service, *Range Plant Handbook*, p. 518.

80. James A. Young, "Tumbleweed," p. 87.

81. Whitson, *Weeds of the West*, notes that the scientific name of *Agropyron repens* has recently been changed to *Elytrigia repens* (p. 411). Darlington discussed the plant as *Triticum repens*, a name that captures its resemblance to wheat or *Triticum* (*American Weeds and Useful Plants*, pp. 390–91), just as the genus *Agropyron* in which it was later included described wheatgrasses. Its common names also include quackgrass, cutchgrass, and quitchgrass.

82. U.S. Department of Agriculture, Forest Service, *Range Plant Handbook*, p. 72.

83. Dewey, "Migration of Weeds," pp. 274–75; Ridley,*Dispersal of Plants*, p. 649.

84. Pammel, *Weeds of the Farm and Garden*, p. 153.

85. Unless otherwise noted, this history of downy brome in the West is based on Mack, "Invasion of *Bromus tectorum*."

86. Muenscher, *Weeds*, p. 147.

87. J. A. Young et al., "Cheatgrass."

88. Ibid., p. 266.

89. The *Range Plant Handbook*, issued by the U.S. Department of Agriculture Forest Service in 1937, explicitly excludes Arizona and New Mexico from the range of downy brome at that time (p. 72). Richard Mack does not address areas farther south than northern California, Nevada, and Utah in "Invasion of *Bromus tectorum*." Forcella and Harvey document a dramatic increase in the range of downy brome in Washington, Oregon, Idaho, Montana, and Wyoming from 1910 to 1930, especially in the 1920s (*New and Exotic Weeds of Montana*, maps, 2:21, graph, 2:101). It is likely that the range of downy brome increased as dramatically elsewhere in the West and that Navajo sheepherders and medicinal experts were familiar with downy brome by the 1930s. The more recent U.S. Department of Agriculture, Agricultural Research Service, *Common Weeds of the United States* (1971), indicates not only that downy brome had been located throughout Arizona and New Mexico but also that some regions had particularly dense concentrations of it (map, p. 46). Regarding Navajo uses of downy brome, see Moerman, *Medicinal Plants of Native America*, 1:97.

90. J. A. Young et al., "Cheatgrass."

91. Regarding johnsongrass, see, for example, Pammel, *Weeds of the Farm and Garden*, p. 139; Whitson, *Weeds of the West*, p. 494; and Spencer, *All About Weeds*, pp. 24–27. Regarding

couchgrass, see U.S. Department of Agriculture, Forest Service, *Range Plant Handbook*, p. 2; Pammel, *Weeds of the Farm and Garden*, pp. 153–55; Georgia, *Manual of Weeds*, p. 63; and Whitson, *Weeds of the West*, p. 411.

92. Spencer, *All About Weeds*, p. 26.

93. Piper, *Forage Plants*, p. 337.

94. Ibid., p. 249.

95. Georgia, *Manual of Weeds*, pp. 61–62.

96. Piper, *Forage Plants*, p. 280.

97. Spencer, *All About Weeds*, p. 29. Regarding crabgrass, see also Pammel, *Weeds of the Farm and Garden*, p. 141; Georgia, *Manual of Weeds*, pp. 26–27; and Whitson, *Weeds of the West*, p. 441.

98. Wheeler, *Forage and Pasture Crops*, p. 560.

99. Piper, *Forage Plants*, pp. 248–49.

100. Wheeler, *Forage and Pasture Crops*, p. 560.

101. Piper, *Forage Plants*, p. 248; Wheeler, *Forage and Pasture Crops*, p. 561.

102. Piper, *Forage Plants*, p. 249. Wheeler states that crested wheatgrass can be grazed almost a month earlier than native grasses (*Forage and Pasture Crops*, p. 562).

103. Wheeler, *Forage and Pasture Crops*, p. 562.

104. Ibid., p. 561.

105. Westover and Rogler, "Crested Wheatgrass."

106. Wheeler, *Forage and Pasture Crops*, p. 561.

107. Montgomery, *Weeds of the Northern U.S. and Canada*, p. 9.

108. Phillips and Foy, *Random House Book of Herbs*, p. 140.

109. Ibid.; Haughton, *Green Immigrants*, pp. 51–52.

110. Darlington, *American Weeds and Useful Plants*, p. 225. Darlington and Haughton (*Green Immigrants*, pp. 51–52) both refer to Ransted's introduction of the plant during the pre-Revolutionary period. Whitson, *Weeds of the West*, claims that toadflax was introduced as an ornamental plant in the middle of the nineteenth century, demonstrating the general inconsistency common to plant histories (p. 549). It is, of course, possible that Ransted brought some toadflax to his garden centuries ago and that someone else introduced the plant again more recently.

111. Georgia, *Manual of Weeds*, p. 379.

112. Moerman, *Medicinal Plants of Native America*, 1:262.

113. Forcella and Harvey, *New and Exotic Weeds of Montana*, maps, 2:62.

114. Whitson, *Weeds of the West*, p. 549. It is unlikely that such an invader would have passed unnoticed if it had been widespread in the 1930s, when range condition attracted so much attention. According to Forcella and Harvey, who documented weed migration in the Northwest, downy brome spread more rapidly than yellow toadflax: in four years, toadflax might move into three counties where it had been absent before, whereas downy brome might move into five or more counties. Downy brome was found in about four times as many counties as yellow toadflax by 1930 (*New and Exotic Weeds of Montana*, 2:101, 108). Although Forcella and Harvey's data do not indicate the acreage of plants—which is always difficult to determine—they do suggest that downy brome was the more "aggressive" and widespread of these two plants in the Northwest. U.S. Department of Agriculture, Forest Service, *Range Plant Handbook*, mentions a genus of related plants primarily native to the West—the owlclovers, or *Orthocarpus*—among which there is a variety called butter-and-

eggs. The owlclovers, like toadflax, make poor forage and likewise were at times pressed into use as ornamental plants (p. 464). See also Spellenberg, *Audubon Society Field Guide to North American Wildflowers, Western Region*, p. 764.

115. Pammel writes that kochia was a common escape, especially in the West (*Weeds of the Farm and Garden*, p. 168). U.S. Department of Agriculture, Agricultural Research Service, *Common Weeds of the United States*, notes that kochia is more abundant in the Midwest but common as far west as Idaho and as far south as Nevada, Utah, and Colorado (pp. 136–37). Muenscher describes kochia as most common in the Midwest (*Weeds*, p. 213). Forcella and Harvey indicate that kochia had been found in one county in Wyoming by 1890; it began to spread slowly in the 1920s and more rapidly after 1940 (*New and Exotic Weeds of Montana*, maps, 2:57). Had kochia been common throughout the West by 1920, Forcella and Harvey would probably have had more evidence of its spread in the Northwest by that time.

116. Whitson, *Weeds of the West*, p. 275.

117. Foy, Forney, and Cooley, "History of Weed Introductions," p. 82.

118. "Summarized History of the Nebraska Noxious Weed Law."

119. "South Dakota Noxious Weed Program"; "Summary of the Kansas Noxious Weed Law."

120. Muenscher, *Weeds*, table 10, pp. 34–35. Only four states had not enacted legislation or did not respond to Muenscher's request for information, all of which were in the East: Louisiana, Georgia, New Hampshire, and Maine. Laws enacted to force landowners (or agencies) to destroy weeds growing in fields, on roadsides, or on public lands began to appear in the late nineteenth century as well. These tended to emphasize eradication as the goal of weed control through the 1980s. Not all states adopted them, partly because they can be difficult to enforce. Weed laws rely on the accurate inspection of millions of acres of range and farmland, the expertise of weed inspectors, the cooperation of landowners, and the feasibility of "eradication." It is much simpler to inspect samples of seed grain. See Westbrooks, "Commentary on New Weeds," pp. 232–37.

121. Pammel, *Weeds of the Farm and Garden*, p. 5.

122. Georgia, *Manual of Weeds*, p. 7.

123. Klingman, Ashton, and Noordhoff, *Weed Science*, p. 1.

124. Pammel, *Weeds of the Farm and Garden*, p. 87; Georgia, *Manual of Weeds*, pp. 2–3; Muenscher, *Weeds*, p. 57; Wilfred W. Robbins, Crafts, and Raynor, *Weed Control*, p. 102.

125. See Muenscher, *Weeds*, pp. 57–60.

126. Ibid.

127. This is still the case. The expense of range weed control is one of the reasons Colorado landowners would not tolerate a weed law that banned a dozen weeds; the 1990 law named only four. See also *Leafy Spurge*, p. 4, and Ross, "Range Tips," p. 26.

128. See Wilfred W. Robbins, Crafts, and Raynor, *Weed Control*, pp. 385–95.

129. Brenchley, *Weeds of Farm Land*, p. 175.

130. Wilfred W. Robbins, Crafts, and Raynor, *Weed Control*, p. 148.

131. H. L. Bolley, "Weed Control by Means of Chemical Sprays," North Dakota Agricultural Experiment Station Bulletin #80, 1908, cited in ibid., p. 149.

132. Ibid. Pammel studied chemical weed control in Iowa. At least two other researchers were working in the East, in Rhode Island and Ohio.

133. Ibid., pp. 149, 153–79.

134. Ibid., p. 152.

135. Ibid., pp. 187–90, 193–200.

136. Mitchell, "Plant Growth Regulators," pp. 256–61.

137. Klingman, Ashton, and Noordhoff, *Weed Science*, p. 299.

138. Ibid., p. 82.

139. Mitchell, "Plant Growth Regulators," p. 264.

140. Wilfred W. Robbins, Crafts, and Raynor, *Weed Control*, p. 102.

141. Klingman, Ashton, and Noordhoff, *Weed Science*, p. v.

142. U.S. Department of Agriculture, Agricultural Research Administration, Bureau of Plant Industry, Soils, and Agricultural Engineering, Division of Weed Investigations, *Bibliography of Weed Investigations* (Washington, D.C.: Government Printing Office, 1951). Annual weed investigation bibliographies cited many articles about 2,4-D in the 1950s.

143. Merck & Company worked cooperatively with the New Jersey agricultural experiment station in studying a Trinidadian shrub that was found to have a high toxicity to the European corn borer. In 1948 Merck chemists isolated the toxin and determined that the bark of the shrub was its best commercial source. Whatever use this shrub may have had in Trinidad, the commercial harvest of its bark would certainly have changed Trinidadians' relationship to it. It is also likely that Trinidadian knowledge led American scientists to the special properties of this shrub. Merck was also involved in pharmaceutical research. See Busbey, "Plants That Help Kill Insects," p. 770, and Raper and Benedict, "Drugs of Microbial Action," p. 737.

144. The history of war-related herbicide research before Vietnam is presented in Cecil, *Herbicidal Warfare*, chap. 1.

145. Clements, *Dynamics of Vegetation*, p. 9.

146. Ibid., pp. 111, 279–89.

147. See Horsman, *Race and Manifest Destiny*, pp. 189–207.

148. Advertisement in *Prairie Farmer* 117 (1 September 1945): 18.

149. Pimentel et al., "Environmental and Economic Effects of Reducing Pesticide Use," indicates that crop losses to weeds decreased from 13.8 percent to 8.5 percent between 1942 and 1960 but have been increasing since 1974. Crop losses to insects rose from 7.1 percent in the 1940s to 13 percent in 1974, a rate of crop loss sustained through the 1980s. Crop loss to disease has also increased in the last fifty years. Crop losses to all pests have increased from 31.4 percent in the 1940s to 37 percent in 1989 (table 2, p. 404). Information I received from various state departments of agriculture and county weed inspectors suggests the same thing. In the states whose weed acreage records I examined, including Kansas, North and South Dakota, and Nebraska, weed acreages appeared to be stable where they were not increasing. Weed acreages are at best good estimates, however, and these states have different standards for weed inspection and vary in their level of interest in conducting a "weed survey" of all noxious weeds in the state. Such a survey is an expensive project. Nevertheless, the overwhelming consensus among weed inspectors was that weeds are a serious problem, a perception that seems not to have diminished in a hundred years, in spite of 2,4-D and other chemicals.

150. U.S. Department of Agriculture, Forest Service, *Range Plant Handbook*, p. 358. Alfilaria was probably also introduced in the hair of cattle and the wool of sheep brought west in the nineteenth century, as a crop contaminant, or in ballast on the west coast (see Ridley, *Dispersal of Plants*, p. 601; Foy, Forney, and Cooley, "History of Weed Introductions," p. 77; and Stein, *My Weeds*, p. 156), but the Spanish were responsible for its first introduction, which was deliberate.

151. Darlington, *American Weeds and Useful Plants*, p. 71.

152. Pammel, *Weeds of the Farm and Garden*, pp. 19, 57, 192.

153. U.S. Department of Agriculture, Forest Service, *Range Plant Handbook*, p. 358.

154. Marguerite Noble, *Filaree*, pp. 4, 9. The novel is dedicated "to those pioneer women who survived in a life of suppression."

155. Deleuze and Guattari, *Thousand Plateaus*, p. 6.

156. Ibid., p. 7.

Epilogue

1. Frederick Jackson Turner, "Significance of the Frontier in American History," p. 2.

2. Ibid., p. 5.

3. Ibid., p. 18.

4. Louis M. Hacker, "Sections—or Classes?," in Taylor, *Turner Thesis*, p. 43; reprinted from *The Nation* 137 (26 July 1933): 108–10.

5. Ibid.

6. Benjamin F. Wright, Jr., "Political Institutions and the Frontier," in Taylor, *Turner Thesis*, p. 34; reprinted from *Sources of Culture in the Middle West*, edited by Dixon Ryan Fox (1934).

7. Billington, *Frederick Jackson Turner*, p. 426.

8. Avery Craven, "Frederick Jackson Turner," in Taylor, *Turner Thesis*, p. 83; reprinted from *Marcus W. Jernegan Essays in American Historiography*, edited by William T. Hutchinson (1937).

9. Wright, "Political Institutions," in Taylor, *Turner Thesis*, pp. 41–42.

10. Letter from Turner to Curti, 30 June 1927, quoted in Billington, *Frederick Jackson Turner*, p. 426.

11. Letter from Turner to Carl Becker, 13 February 1926, quoted in Craven, "Frederick Jackson Turner," in Taylor, *Turner Thesis*, p. 84.

12. Letter from Turner to Charles H. Haskins, 15 July 1896, quoted in Billington, *Frederick Jackson Turner*, p. 160.

13. Ibid., pp. 433–35.

14. Ibid. See also ibid., p. 427.

15. McClure, "Late Imperial Romance."

16. Strong, *Our Country*, pp. 160–61.

17. David W. Noble, *End of American History*, pp. 16, 27.

18. Richards, *Imperial Archive*.

19. Nabhan, *Enduring Seeds*.

Bibliography

Alexander, Thomas G. "From Rule of Thumb to Scientific Range Management: The Case of the Intermountain Region of the Forest Service." *Western Historical Quarterly* 18 (October 1987): 409–28.

Allen, L. F. "The Shorthorn Breed of Cattle." In U.S. Department of Agriculture, *[Annual] Report, 1875*, pp. 416–26. Washington, D.C.: Government Printing Office, 1876.

Allen, Shirley. *An Introduction to American Forestry*. New York: McGraw-Hill, 1950.

Ardrey, Robert. *American Agricultural Implements*. 1894. Reprint, New York: Arno Press, 1972.

Armsby, Henry Prentiss. *The Nutrition of Farm Animals*. New York: Macmillan, 1917.

Atkinson, Alfred. "Dry Farming Investigations in Montana." Montana Agricultural Experiment Station Bulletin #74, December 1908.

Atkinson, Alfred, and F. S. Cooley. "Dry Farming Practice in Montana." Montana Agricultural College Experiment Station Circular #3, February 1910.

Baker, F. S. "Silviculture." In *Fifty Years of Forestry in the U.S.A.*, edited by Robert K. Winters, pp. 65–77. Washington, D.C.: Society of American Foresters, 1950.

Barnes, Will C. *Western Grazing Grounds and Forest Ranges: A History of the Live-Stock Industry on the Open Ranges of the Arid West*. Chicago: Breeders Gazette, 1913.

Beal, W. H. "The Farmer's Vegetable Garden." In "Experiment Station Work," edited by W. H. Beal. U.S. Department of Agriculture Farmers' Bulletin #149, 1902.

Beattie, J. H., and W. R. Beattie. "The Farm Garden." U.S. Department of Agriculture Farmers' Bulletin #1673, 1931.

Beattie, W. R. "The Home Vegetable Garden." U.S. Department of Agriculture Farmers' Bulletin #255, 1906.

Bergoffen, W. W. "Questions and Answers." In U.S. Department of Agriculture, *Trees: The Yearbook of Agriculture, 1949*, pp. 19–36. Washington, D.C.: Government Printing Office, 1949.

Bergson, Henri. *Matter and Memory*. London: S. Sonnenschein, 1911.

Billington, Ray Allen. *Frederick Jackson Turner: Historian, Scholar, Teacher*. New York: Oxford University Press, 1972.

——. *The Westward Movement in the United States*. Princeton, N.J.: D. Van Nostrand, 1959.

Blegen, Theodore. *Building Minnesota*. Boston: D. C. Heath, 1938.

Bloch, Marc. *Feudal Society*. 2 vols. Chicago: University of Chicago Press, 1961.

Blunn, Cecil. "Improvement of the Navajo Sheep." *Journal of Heredity* 31 (March 1940): 104.

Boardman, Samuel P. "Sheep Husbandry in the West." In *Report of the Commissioner of Agriculture, 1862*, edited by Isaac Newton, p. 288. Washington, D.C.: Government Printing Office, 1863.

Bookchin, Murray. *The Ecology of Freedom: The Emergence and Dissolution of Hierarchy*. 1982. Reprint, Montreal: Black Rose Press, 1991.

Braudel, Fernand. *Capitalism and Material Life, 1400–1800*. Translated by Miriam Kochan. San Francisco: Harper & Row, 1973.

Brenchley, Winifred E. *Weeds of Farm Land*. London: Longmans, Green, 1920.
Briggs, Hilton M., and Dinus M. Briggs. *Modern Breeds of Livestock*. 4th ed. New York: Macmillan, 1980.
Brown, Lauren. *Grasslands*. The Audubon Society Nature Guides. New York: Knopf, 1985.
Brown, Nelson. "War and Wood in Germany." *American Forests* 45 (December 1939): 587–623.
Busbey, Ruth L. "Plants That Help Kill Insects." In U.S. Department of Agriculture, *Crops in Peace and War: Yearbook of Agriculture, 1950–1951*, pp. 765–71. Washington, D.C.: Government Printing Office, [1952].
Butler, Ovid. "Editor's Log." *American Forests* 46 (June 1940): 245.
——. "Editor's Log." *American Forests* 47 (March 1941): 101.
——. "Editor's Log." *American Forests* 47 (May 1941): 213.
——. "Editor's Log." *American Forests* 47 (September 1941): 405.
——. "Fire and Defense." *American Forests* 47 (June 1941): 281.
——. "Forests and National Defense." *American Forests* 46 (January 1940): 27.
——. "Forests in the National Defense." *American Forests* 46 (August 1940): 343.
——. "Lumber on the Carpet." *American Forests* 47 (March 1941): 121.
——. "Our Scorched Earth Practice." *American Forests* 48 (May 1941): 407.
——. "The War and the Parks." *American Forests* 49 (March 1943): 121.
Campbell, Robert S. "Milestones in Range Management." *Journal of Range Management* 1 (October 1948): 4–8.
Carpenter, Ronald H. *The Eloquence of Frederick Jackson Turner*. San Marino, Calif.: Huntington Library, 1983.
Castetter, Edward F., and Willis H. Bell. *Pima and Papago Indian Agriculture*. Albuquerque: University of New Mexico Press, 1942.
Cecil, Paul Frederick. *Herbicidal Warfare: The Ranch Hand Project in Vietnam*. New York: Praeger, 1986.
Chase, L. W. "Farm Machinery." In *Fundamentals of Agriculture*, edited by James Halligan, pp. 439–41. Boston: D. C. Heath, 1911.
Civilian Conservation Corps, Federal Security Administration, Soil Conservation Service, and U.S. Forest Service. *The CCC at Work*. Washington, D.C.: Government Printing Office, 1941.
Clapp, Earle. "The Major Range Problems and Their Solution." In *The Western Range: Letter from the Secretary of Agriculture*, edited by H. H. Wallace, pp. 1–69. Washington, D.C.: Government Printing Office, 1936.
Clarke, William James. *Modern Sheep: Breeds and Management*. Chicago: American Sheep Breeder, 1907.
Clastres, Pierre. *Society against the State*. New York: Zone Books, 1987.
Clements, Frederic. *Dynamics of Vegetation*. Compiled and edited by B. W. Allred and Edith S. Clements. New York: H. W. Wilson, 1949.
Clepper, Henry. *Professional Forestry in the United States*. Baltimore: Johns Hopkins University Press, 1971.
Cline, A. C. "Future Requirements for Timber." In U.S. Department of Agriculture, *Trees: The Yearbook of Agriculture, 1949*, pp. 731–41. Washington, D.C.: Government Printing Office, 1949.
Coburn, F. D. *The Book of Alfalfa*. New York: Orange Judd, 1908.

Cochrane, Timothy. "Trial by Fire: Early Forest Service Rangers' Fire Stories." *Forest and Conservation History* 35 (January 1991): 16–23.

Cochrane, Willard. *The Development of American Agriculture: A Historical Analysis.* 2d ed. Minneapolis: University of Minnesota Press, 1993.

Coffey, Joel. *Livestock Management.* Chicago: J. B. Lippincott, 1940.

Cohen, Stan. *The Tree Army: A Pictorial History of the Civilian Conservation Corps, 1933–1942.* Missoula, Mont.: Pictorial Histories Publishing Company, 1980.

Cohen, Stan, and Don Miller. *The Big Burn: The Northwest's Forest Fire of 1910.* Missoula, Mont.: Pictorial Histories Publishing Company, 1978.

Coleman, Donald. "Wood in Modern Warfare." *American Forests* 46 (January 1940): 6–48.

Coleman, Norman. *Report of the Commissioner of Agriculture, 1887.* Washington, D.C.: Government Printing Office, 1888.

"The Collection History of *Centaureas.*" Washington State University Research Bulletin XB 0978, 1980.

"The Conservation War Front." *American Forests* 48 (March 1942): 109.

Coon, Nelson. *Using Plants for Healing.* Illustrated by Kenneth Ranier. Emmaus, Pa.: Rodale Press, 1979.

Corbett, Lee Cleveland. *Garden Farming.* Boston: Ginn, 1913.

Coromines, Joan. *Diccionari Etimològic i Complementari de la Llengua Catalonia.* Barcelona: Curial Editions Catalanes Caixa de Pensions "La Caixa," 1988.

Coville, Frederick V. "Bureau of Plant Industry Report: Taxonomic and Range Investigations." In U.S. Department of Agriculture, *Annual Reports of the Department of Agriculture, 1910*, pp. 304–7. Washington, D.C.: Government Printing Office, 1911.

Cowan, Charles S. *The Enemy Is Fire!* Seattle: Superior Publishing Company, 1961.

Cox, William T. "Wild Rubber: Can We Get It?" *American Forests* 48 (April 1942): 170–89.

Cronon, William. "Modes of Prophecy and Production: Placing Nature in History." *Journal of American History* 76 (March 1990): 1122–31.

——. *Nature's Metropolis: Chicago and the Great West.* New York: W. W. Norton, 1991.

——. "Revisiting the Vanishing Frontier: The Legacy of Frederick Jackson Turner." *Western Historical Quarterly* 18 (April 1987): 157–76.

Crosby, Alfred. *Ecological Imperialism: The Biological Expansion of Europe, 900–1900.* Cambridge: Cambridge University Press, 1986.

——. "An Enthusiastic Second." *Journal of American History* 76 (March 1990): 1107–10.

Danbom, David. *The Resisted Revolution: Urban America and the Industrialization of Agriculture, 1900–1930.* Ames: Iowa State University Press, 1979.

Darlington, William. *American Weeds and Useful Plants, Being a Second and Illustrated Edition of Agricultural Botany: An Enumeration and Description of Useful Plants and Weeds, Which Merit the Notice, or Require the Attention of American Agriculturists.* 2d ed., rev. and enl. 1859. Reprint, New York: Orange Judd, 1883.

Deleuze, Gilles, and Félix Guattari. *A Thousand Plateaus: Capitalism and Schizophrenia.* Minneapolis: University of Minnesota Press, 1987.

Deloria, Vine, Jr. "Circling the Same Old Rock." In *Marxism and Native Americans*, edited by Ward Churchill, pp. 113–36. Boston: South End Press, 1982.

Denning, Michael. " 'The Special American Conditions': Marxism and American Studies." *American Quarterly* 38 (Fall 1986): 356–80.

Derrida, Jacques. *Margins of Philosophy*. Translated by Alan Bass. Chicago: University of Chicago Press, 1982.

Dewey, Lyster H. "Canada Thistle." U.S. Department of Agriculture Bureau of Botany Circular #27, 1901.

——. "Migration of Weeds." U.S. Department of Agriculture, *Yearbook of the Department of Agriculture, 1896*, pp. 263–86. Washington, D.C.: Government Printing Office, 1897.

Drinnon, Richard. *Facing West: The Metaphysics of Indian-Hating and Empire-Building*. 1980. Reprint, New York: Schocken Books, 1990.

Dyer, Christopher. "Documentary Evidence: Problems and Enquiries." In *The Countryside of Medieval England*, edited by Grenville Astill and Annie Grant, pp. 12–35. Oxford: Basil Blackwell, 1988.

Edwards, Everett E. "The Settlement of the Grasslands." In U.S. Department of Agriculture, *Grass: Yearbook of Agriculture, 1948*, pp. 16–34. Washington, D.C.: Government Printing Office, 1948.

Ensminger, M. E. *Sheep Husbandry*. Danville, Ill.: Interstate Press, 1952.

Faulkner, Edward. *Plowman's Folly and A Second Look*. 1943, 1947. Reprint, with an introduction by Charles Little, Washington, D.C.: Island Press, 1987.

Fernow, Bernhard. *A Brief History of Forestry in Europe, the United States, and Other Countries*. Toronto: University of Toronto Press, 1913.

——. *Economics of Forestry: A Reference Book for Students of Political Economy and Professional and Lay Students of Forestry*. New York: Thomas Y. Crowell, 1902.

"First Spruce Raft Arrives from Alaska." *American Forests* 49 (March 1943): 132.

Fite, Gilbert. *The Farmers' Frontier, 1865–1900*. New York: Holt, Rinehart and Winston, 1966.

Fletcher, R. A., and A. J. Renney. "A Growth Inhibitor Found in *Centaurea* spp." *Canadian Journal of Plant Science* 43 (1963): 475–81.

Forcella, Frank, and Stephen J. Harvey. *New and Exotic Weeds of Montana*. 2 vols. Bozeman: Montana State University Herbarium, 1980–81.

"Forestry in Congress." *American Forests* 46 (November 1940): 520–21.

Foucault, Michel. *The Order of Things: An Archaeology of the Human Sciences*. New York: Vintage Books, 1973.

Foy, C. L., D. R. Forney, and W. E. Cooley. "History of Weed Introductions." In *Exotic Plant Pests and North American Agriculture*, edited by Charles L. Wilson and Charles L. Graham, pp. 65–92. New York: Academic Press, 1983.

Frye, Northrop. *Anatomy of Criticism: Four Essays*. Princeton: Princeton University Press, 1957.

Funk, W. C. "What the Farm Contributes Directly to the Farmer's Living." U.S. Department of Agriculture Farmers' Bulletin #635, 1914.

Fussell, Betty. *The Story of Corn*. New York: Knopf, 1992.

Gamboa, Erasmo. *Mexican Labor and World War II: Braceros in the Pacific Northwest, 1942–1947*. Austin: University of Texas Press, 1990.

Georgia, Ada. *Manual of Weeds*. 1914. Reprint, New York: Macmillan, 1930.

Gonzales, Juan L., Jr. *Mexican and Mexican American Farm Workers: The California Agricultural Industry*. New York: Praeger, 1985.

Graff, Gerald. *Professing Literature: An Institutional History*. Chicago: University of Chicago Press, 1987.

Granger, C. M. "The National Forests at War." *American Forests* 49 (March 1943): 112–38.

Greathouse, Charles. "The Vegetable Garden." U.S. Department of Agriculture Farmers' Bulletin #94, 1899.
Greeley, William. *Forest Policy*. New York: McGraw-Hill, 1953.
——. *Forests and Men*. Garden City, N.Y.: Doubleday, 1951.
Green, William. "The Real Interest of the People." In U.S. Department of Agriculture, *Trees: The Yearbook of Agriculture, 1949*, pp. 754–55. Washington, D.C.: Government Printing Office, 1949.
Grieg, James. "Plant Resources." In *The Countryside of Medieval England*, edited by Grenville Astill and Annie Grant, pp. 107–27. Oxford: Basil Blackwell, 1988.
Griffiths, Francis P., Harold S. Olcott, and W. Lawrence Shaw. "Our Second Largest Food Group." In U.S. Department of Agriculture, *Crops in Peace and War: Yearbook of Agriculture, 1950–51*, pp. 213–16. Washington, D.C.: Government Printing Office, [1952].
"Haiti to Plant Rubber for U.S." *American Forests* 47 (September 1941): 446.
Hallyn, Fernand. *The Poetic Structure of the World: Copernicus and Kepler*. New York: Zone Books, 1990.
Hammer, Kenneth M. "Bonanza Farming: Forerunner of Modern Large-Scale Farming." In *Agriculture in the West*, edited by Edward L. Schapsmeier and Frederick H. Schapsmeier, pp. 52–61. Manhattan, Kans.: Sunflower University Press, 1980.
Hargreaves, Mary Wilma M. *Dry Farming in the Northern Great Plains, 1900–1925*. Cambridge: Harvard University Press, 1957.
Harrington, H. D. *Edible Native Plants of the Rocky Mountains*. Illustrated by Y. Matsumura. Albuquerque: University of New Mexico Press, 1967.
Harrison, Robert Pogue. *Forests: The Shadow of Civilization*. Chicago: University of Chicago Press, 1992.
Haughton, Claire Shaver. *Green Immigrants: The Plants That Transformed America*. New York: Harcourt Brace Jovanovich, 1978.
Hay, Howard. "For CCC Military Training." *American Forests* 45 (April 1939): 242.
Hays, Samuel P. *Conservation and the Gospel of Efficiency: The Progressive Conservation Movement, 1890–1920*. Cambridge: Harvard University Press, 1959.
Henius, Frank. "Oriente: New Empire of the Americas." *American Forests* 48 (May 1942): 391–426.
Henry, W. A., and F. B. Morrison. *Feeds and Feeding: A Handbook for the Student and Stockman*. Rev. ed. Madison, Wis.: Henry-Morrison Company, 1915.
Hilgard, E. W., and W. J. V. Osterhout. *Agriculture for Schools of the Pacific Slope*. New York: Macmillan, 1910.
Hitchcock, A. S. *Manual of the Grasses of the United States*. 2d ed. U.S. Department of Agriculture Miscellaneous Publication #200. Washington, D.C.: Government Printing Office, 1950.
Holbrook, Stewart Hall. *Burning an Empire: The Story of American Forest Fires*. New York: Macmillan, 1943.
——. "The Forest Goes to War." *American Forests* 48 (February 1942): 55–62.
——. "Forty Men and a Fire." *American Forests* 46 (June 1940): 251–53.
——. *Machines of Plenty: Pioneering in American Agriculture*. New York: Macmillan, 1955.
Holder, Preston. *The Hoe and the Horse on the Plains: A Study of Cultural Development among North American Indians*. 1970. Reprint, Lincoln: University of Nebraska Press, 1991.

Hoover, Max M., M. A. Hein, William A. Dayton, and C. O. Erlanson. "The Main Grasses for Farm and Home." In U.S. Department of Agriculture, *Grass: Yearbook of Agriculture, 1948*, pp. 639–700. Washington, D.C.: Government Printing Office, 1948.

Horlacher, Levi Jackson, and Carsie Hammonds. *Sheep*. Danville, Ill.: Interstate Press, 1953.

Horsman, Reginald. *Race and Manifest Destiny*. Cambridge: Harvard University Press, 1981.

Hunt, George M. "The Forest Products Laboratory." In U.S. Department of Agriculture, *Trees: The Yearbook of Agriculture, 1949*, pp. 647–51. Washington, D.C.: Government Printing Office, 1949.

Hunt, John Clark. "Fire-Fighting after the War." *American Forests* 49 (October 1943): 470–73.

——. "If War Comes to the Forest." *American Forests* 47 (September 1941): 407–47.

Hurt, R. Douglas. *Agricultural Technology in the Twentieth Century*. Manhattan, Kans.: Sunflower University Press, 1991.

——. *American Farm Tools from Hand-Power to Steam-Power*. Manhattan, Kans.: Sunflower University Press, 1982.

——. *The Dust Bowl: An Agricultural and Social History*. Chicago: Nelson-Hall, 1981.

——. *Indian Agriculture in America, Prehistory to the Present*. Lawrence: University Press of Kansas, 1987.

——. "The National Grasslands: Origin and Development in the Dust Bowl." *Agricultural History* 59 (April 1985): 246–59.

The Implement Blue Book. St. Louis: Midland Publishing Company, 1906.

"Incendiary Fires Sweep California." *American Forests* 47 (November 1941): 535.

Ingalls, John James. "In Praise of Blue Grass." *Kansas Magazine*, 1872. Reprinted in U.S. Department of Agriculture, *Grass: The Yearbook of Agriculture, 1948*, pp. 6–8. Washington, D.C.: Government Printing Office, 1948.

Ise, John. *The United States Forest Policy*. New Haven: Yale University Press, 1920.

Ives, Judge. "How the Russians Plow." *Farm Implements* 13 (27 May 1899): 15.

Jaimes, M. Annette. "Re-Visioning Native America: An Indigenist View of Primitivism and Industrialism." *Social Justice* 19 (Summer 1992): 5–34.

Jehlen, Myra, and Sacvan Bercovitch. *Ideology and Classic American Literature*. Cambridge: Cambridge University Press, 1986.

Jenkins, Craig J. *The Politics of Insurgency: The Farm Worker Movement of the 1960s*. New York: Columbia University Press, 1985.

Johnson, Paul C. *Farm Animals in the Making of America*. Des Moines: Wallace Homestead Book Company, 1975.

Jones, Henry Albert, and Joseph Tooker Rosa. *Truck Crop Plants*. New York: McGraw-Hill, 1928.

Jones, Martin. "Regional Patterns in Crop Production." In *Aspects of the Iron Age in Central Southern Britain*, edited by Barry Cunliffe and David Miles, pp. 120–25. Oxford: Oxford University Committee for Archaeology, 1984.

Jones, Pamela. *Just Weeds: History, Myths, and Uses*. New York: Prentice Hall Press, 1991.

Katz, William Loren. *The Black West*. Seattle: Open Hand Publishing, 1987.

Kauffman, Erle. "Guayule . . . the 'Victory' Rubber." *American Forests* 48 (February 1942): 72–92.

——. "Spruce Goes Back to War." *American Forests* 46 (August 1940): 363–64.
Keffer, Charles. "The Garden." In *Fundamentals of Agriculture*, edited by James Halligan, pp. 232–34. Boston: D. C. Heath, 1911.
Keller, Wesley, and H. R. Hochmuth. "Cultivated Forage Crops." In U.S. Department of Agriculture, *Grass: Yearbook of Agriculture, 1948*, pp. 548–53. Washington, D.C.: Government Printing Office, 1948.
Kelly, Lawrence C. "Anthropology in the Soil Conservation Service." *Agricultural History* 59 (April 1985): 136–47.
Kerr, Edmund. "Sky-Fighters of the Forest." *American Forests* 49 (September 1943): 431–62.
Klingman, Glenn C., Floyd M. Ashton, and Lyman J. Noordhoff. *Weed Science: Principles and Practices*. 2d ed. New York: John Wiley & Sons, 1982.
Kolodny, Annette. *The Land Before Her: Fantasy and Experience of the American Frontiers, 1630–1860*. Chapel Hill: University of North Carolina Press, 1984.
——. *The Lay of the Land: Metaphor as Experience and History in American Life and Letters*. Chapel Hill: University of North Carolina Press, 1975.
Kraenzel, Carl Frederick. *The Great Plains in Transition*. Norman: University of Oklahoma Press, 1955.
Kuklick, Bruce. "Myth and Symbol in American Studies." *American Quarterly* 24 (October 1972): 435–50.
Lacey, Celestine A., Charles Egan, Wayne Pearson, and Peter K. Fay. "Bounty Programs: An Effective Management Tool." *Weed Technology* 2 (April 1988): 196–97.
Leafy Spurge Symposium Proceedings. Bozeman: Montana State University, June 1982.
Levenstein, Harvey. *Paradox of Plenty: A Social History of Eating in Modern America*. New York: Oxford University Press, 1993.
——. *Revolution at the Table: The Transformation of the American Diet*. New York: Oxford University Press, 1988.
Lévy-Bruhl, Lucien. *How Natives Think*. New York: Washington Square Press, 1966.
Lewis, R. W. B. *The American Adam: Innocence, Tragedy, and Tradition in the Nineteenth Century*. Chicago: University of Chicago Press, 1955.
Limerick, Patricia Nelson. *The Legacy of Conquest: The Unbroken Past of the American West*. New York: W. W. Norton, 1987.
Linfield, F. B. "Dryland Farming in Montana." Montana Agricultural Experiment Station Bulletin #63, January 1907.
McAlister, Lyle N. *Spain and Portugal in the New World, 1492–1700*. Minneapolis: University of Minnesota Press, 1984.
McClure, John. "Late Imperial Romance." *Raritan* 10 (Spring 1991): 111–30.
McEntee, James. *Now They Are Men*. Washington, D.C.: National Home Library Foundation, 1940.
Mack, Richard. "Invasion of *Bromus tectorum* L. into Western North America: An Ecological Chronicle." *Agro-Ecosystems* 7 (1981): 145–65.
McKee, Roland. "The Legumes of Many Uses." In U.S. Department of Agriculture, *Grass: Yearbook of Agriculture, 1948*, pp. 701–26. Washington, D.C.: Government Printing Office, 1948.
MacLean, Norman. *Young Men and Fire*. Chicago: University of Chicago Press, 1992.
Malin, James C. *History and Ecology: Studies of the Grassland*. Edited by Robert Swierenga. Lincoln: University of Nebraska Press, 1984.

Mann, Stuart E. *An Indo-European Comparative Dictionary*. Hamburg: Helmut Buske Verlag, 1984/87.

Marcus, Alan. *Agricultural Science and the Quest for Legitimacy: Farmers, Agricultural Colleges, and Experiment Stations, 1870–1890*. Ames: Iowa State University Press, 1985.

Marks, Barry. "The Concept of Myth in *Virgin Land*." *American Quarterly* 5 (Spring 1953): 71–76.

Marsh, George Perkins. *Man and Nature: Or, Physical Geography as Modified by Human Action*. Edited by David Lowenthal. 1864. Reprint, Cambridge: Harvard University Press, 1965.

Marx, Leo. *The Machine in the Garden: Technology and the Pastoral Ideal in America*. New York: Oxford University Press, 1964.

——. *The Pilot and the Passenger: Essays on Literature, Technology, and Culture in the United States*. New York: Oxford University Press, 1988.

Matthiessen, F. O. *American Renaissance: Art and Expression in the Age of Emerson and Whitman*. New York: Oxford University Press, 1941.

Merchant, Carolyn. *The Death of Nature: Women, Ecology, and the Scientific Revolution*. 1980. Reprint, San Francisco: Harper & Row, 1989.

——. *Ecological Revolutions: Nature, Gender, and Science in New England*. Chapel Hill: University of North Carolina Press, 1989.

——. "Gender and Environmental History." *Journal of American History* 76 (March 1990): 1117–21.

Mies, Maria. *Patriarchy and Accumulation on a World Scale: Women in the International Division of Labor*. London: Zed Books, 1987.

Mies, Maria, and Vandana Shiva. *Ecofeminism*. London: Zed Books, 1993.

Miller, Perry. *Errand into the Wilderness*. Cambridge: Harvard University Press, 1964.

Mitchell, John W. "Plant Growth Regulators." In U.S. Department of Agriculture, *Science in Farming: Yearbook of Agriculture, 1943–1947*, pp. 256–61. Washington, D.C.: Government Printing Office, 1947.

Moerman, Daniel E. *Medicinal Plants of Native America*. 2 vols. University of Michigan Museum of Anthropology, Technical Reports #19, and Research Reports in Ethnobotany, Contribution #2. Ann Arbor: University of Michigan Museum of Anthropology, 1986.

Montgomery, F. H. *Weeds of the Northern U.S. and Canada*. New York: Frederick Warne, 1964.

Morgan, Captain A. S. "Greater America." *Farm Implements* 13 (28 January 1899): 23–25.

Morgan, Dan. *Merchants of Grain*. New York: Viking, 1979.

Muenscher, Walter Conrad. *Weeds*. 2d ed. New York: Macmillan, 1955.

Murray, Philip. "Labor Looks at Trees and Conservation." In U.S. Department of Agriculture, *Trees: The Yearbook of Agriculture, 1949*, pp. 755–57. Washington, D.C.: Government Printing Office, 1949.

Nabhan, Gary. *Enduring Seeds: Native American Agriculture and Wild Plant Conservation*. San Francisco: North Point Press, 1989.

"National Defense and Public Regulations Feature Forester's Annual Report." *American Forests* 47 (March 1941): 126.

"National Defense Lays Heavy Lumber Demand upon Industry." *American Forests* 46 (October 1940): 475.

"Need for Cut in Lumber Uses Weighed by WPB [War Production Board]." *American Forests* 50 (May 1944): 248.
Nelson, Oran M. "Sheep." In *Western Live-Stock Management*, edited by Ermine L. Potter, pp. 119–239. New York: Macmillan, 1917.
Newton, Isaac. *Report of the Commissioner of Agriculture, 1862*. Washington, D.C.: Government Printing Office, 1863.
Nietzsche, Friedrich. *Philosophy and Truth: Selections from Nietzsche's Notebooks of the Early 1870s*. Edited and translated by Daniel Breazeale. Atlantic Highlands, N.J.: Humanities Press International, 1979.
Noble, David W. *The End of American History: Democracy, Capitalism, and the Metaphor of Two Worlds in Anglo-American Historical Writing, 1880–1980*. Minneapolis: University of Minnesota Press, 1985.
Noble, Marguerite. *Filaree*. 1979. Reprint, Albuquerque: University of New Mexico Press, 1985.
Noriega, Jorge. "American Indian Education in the United States." In *The State of Native America: Genocide, Colonization, and Resistance*, edited by M. Annette Jaimes, pp. 371–402. Boston: South End Press, 1992.
Norwood, Vera. *Made From This Earth: American Women and Nature*. Chapel Hill: University of North Carolina Press, 1993.
"On the Forest Fire Front." *American Forests* 48 (May 1942): 247–84.
Osgood, Ernest Staples. *The Day of the Cattleman*. 1929. Reprint, Chicago: University of Chicago Press, 1964.
Oxford English Dictionary. Oxford: Oxford University Press, 1971.
Pammel, Louis Hermann. *Weeds of the Farm and Garden*. New York: Orange Judd, 1911.
Parman, Donald L. *The Navajos and the New Deal*. New Haven: Yale University Press, 1976.
Pechanec, Joseph. "Our Range Society." *Journal of Range Management* 1 (October 1948): 1.
Phillips, Roger, and Nicky Foy. *The Random House Book of Herbs*. New York: Random House, 1990.
Pimentel, David, et al. "Environmental and Economic Effects of Reducing Pesticide Use." *BioScience* 41 (June 1991): 402–9.
Pinchot, Gifford. *Breaking New Ground*. New York: Harcourt Brace, 1947.
——. *A Primer of Forestry*. Washington, D.C.: Government Printing Office, 1905.
Piper, Charles V. *Forage Plants and Their Culture*. Rev. ed. New York: Macmillan, 1924.
Pisani, Donald. *From the Family Farm to Agribusiness: The Irrigation Crusade in California and the West, 1850–1931*. Berkeley: University of California Press, 1984.
——. *To Reclaim a Divided West: Water, Law, and Public Policy, 1848–1902*. Albuquerque: University of New Mexico Press, 1992.
Potter, Ermine L. *Western Live-Stock Management*. New York: Macmillan, 1917.
Pyne, Stephen. *Fire in America: A Cultural History of Wildland and Rural Fire*. Princeton: Princeton University Press, 1988.
——. "Firestick History." *Journal of American History* 76 (March 1990): 1132–41.
Raper, Kenneth B., and Robert Benedict. "The Drugs of Microbial Action." In U.S. Department of Agriculture, *Crops in Peace and War: Yearbook of Agriculture, 1950–1951*, pp. 734–41. Washington, D.C.: Government Printing Office, [1952].
Record, Samuel, and Robert Hess. *Timbers of the New World*. New Haven: Yale University Press, 1943.

"Regulation Urgent, Says Forest Service." *American Forests* 48 (March 1942): 134.

Reising, Russell. *The Unusable Past: Theory and the Study of American Literature*. New York: Methuen, 1986.

Richards, Thomas. *The Imperial Archive: Knowledge and the Fantasy of Empire*. London: Verso, 1993.

Ridley, Henry N. *The Dispersal of Plants throughout the World*. Ashford, Eng.: L. Reeve, 1930.

Robbins, Wilfred W., Alden Crafts, and Richard Raynor. *Weed Control: A Textbook and Manual*. New York: McGraw-Hill, 1942.

Robbins, William G. *American Forestry: A History of National, State, and Private Cooperation*. Lincoln: University of Nebraska Press, 1985.

——. *Hard Times in Paradise: Coos Bay, Oregon, 1850–1986*. Seattle: University of Washington Press, 1988.

Roessel, Ruth, and Broderick H. Johnson, comp. *Navajo Livestock Reduction: A National Disgrace*. Chinle, Ariz.: Navajo Community College Press, 1974.

Rogin, Leo. *The Introduction of Farm Machinery in Its Relation to the Productivity of Labor in the Agriculture of the United States during the Nineteenth Century*. Berkeley: University of California Press, 1931.

Rosaldo, Renato. *Culture and Truth: The Remaking of Social Analysis*. Boston: Beacon Press, 1989.

Ross, Rob. "Range Tips." Montana State University Cooperative Extension Service, Bulletin #1307, May 1984.

Rowley, William D. *U.S. Forest Service Grazing and Rangelands: A History*. College Station: Texas A&M University Press, 1985.

Ruiz, Vicki L. *Cannery Women, Cannery Lives: Mexican Women, Unionization, and the California Food Processing Industry, 1930–1950*. Albuquerque: University of New Mexico Press, 1987.

Sachs, Carolyn. *The Invisible Farmers: Women in Agricultural Production*. Totowa, N.J.: Rowman & Allenheld, 1983.

Sahlins, Marshall. *Stone Age Economics*. New York: Aldine, 1972.

Salmond, John. *The Civilian Conservation Corps, 1933–1942: A New Deal Case Study*. Durham: Duke University Press, 1967.

Sampson, Arthur. *Range and Pasture Management*. New York: John Wiley & Sons, 1923.

——. *Range Management: Principles and Practices*. New York: John Wiley & Sons, 1952.

——. "Succession as a Factor in Range Management." *Journal of Forestry* 15 (May 1917): 593–96.

——. "Suggestions for Instruction in Range Management." *Journal of Forestry* 17 (May 1919): 523–45.

Sandoz, Mari. *The Cattlemen from the Rio Grande across the Far Marias*. 1958. Reprint, Lincoln: University of Nebraska Press, 1978.

Savage, D. A., James E. Smith, and D. F. Costello. "Dry-Land Pastures on the Plains." In U.S. Department of Agriculture, *Grass: The Yearbook of Agriculture, 1948*, pp. 506–22. Washington, D.C.: Government Printing Office, 1948.

Schilletter, Julian Claude, and Harry Wyatt Richey. *Textbook of General Horticulture*. New York: McGraw-Hill, 1940.

Shaw, Thomas. *The Study of Breeds in America*. New York: Orange Judd, 1912.

Shiva, Vandana. *Monocultures of the Mind: Perspectives on Biodiversity and Biotechnology*. London: Zed Books, 1993.
———. *The Violence of the Green Revolution: Third World Agriculture, Ecology, and Politics*. London: Zed Books, 1991.
Silko, Leslie. *Ceremony*. New York: Viking, 1977.
Simmons, Fred. "Yesterday and Today: Since the Days of Leif Ericson." In U.S. Department of Agriculture, *Trees: The Yearbook of Agriculture, 1949*, pp. 687–94. Washington, D.C.: Government Printing Office, 1949.
Slotkin, Richard. *The Fatal Environment: The Myth of the Frontier in the Age of Industrialization, 1800–1890*. Middletown, Conn.: Wesleyan University Press, 1985.
Smith, Henry Nash. "Symbol and Idea in *Virgin Land*." In *Ideology and Classic American Literature*, edited by Myra Jehlen and Sacvan Bercovitch, pp. 21–35. Cambridge: Cambridge University Press, 1986.
———. *Virgin Land: The American West as Symbol and Myth*. 1950. Reprint, Cambridge: Harvard University Press, 1970.
Solbrig, Otto T., and Dorothy J. Solbrig. *So Shall You Reap: Farming and Crops in Human Affairs*. Washington, D.C.: Island Press, 1994.
"South Dakota Noxious Weed Program: In the Beginning." Prepared by the South Dakota Department of Agriculture, 1991.
Sparhawk, W. N. "The History of Forestry in America." In U.S. Department of Agriculture, *Trees: The Yearbook of Agriculture, 1949*, pp. 702–14. Washington, D.C.: Government Printing Office, 1949.
Spellenberg, Richard. *The Audubon Society Field Guide to North American Wildflowers, Western Region*. New York: Knopf, 1979.
Spencer, Edwin Rollin. *All About Weeds*. 1957. Reprint, New York: Dover, 1974.
Spillman, W. J. "Grass and Forage Plant Investigations." In U.S. Department of Agriculture, *Annual Reports of the Department of Agriculture, 1905*, pp. 111–28. Washington, D.C.: Government Printing Office, 1906.
Stein, Sara B. *My Weeds: A Gardener's Botany*. New York: Harper & Row, 1988.
Strong, Josiah. *Our Country: Its Possible Future and Its Present Crisis*. New York: Baker & Taylor, 1885.
"A Summarized History of the Nebraska Noxious Weed Law." Prepared by the Nebraska Department of Agriculture, [1989].
"A Summary of the Kansas Noxious Weed Law." Prepared by the Kansas State Board of Agriculture, 1990.
Talbot, M. W., and E. C. Crafts. "The Lag in Research and Extension." In U.S. Department of Agriculture, Forest Service, *The Western Range: Letter from the Secretary of Agriculture*, edited by H. A. Wallace, pp. 185–92. Washington, D.C.: Government Printing Office, 1936.
Tannahill, Reay. *Food in History*. Rev. ed. New York: Crown Publishers, 1988.
Tate, Cecil. *The Search for a Method in American Studies*. Minneapolis: University of Minnesota Press, 1973.
Taylor, George Roger, ed. *The Turner Thesis Concerning the Role of the Frontier in American History*. Boston: D. C. Heath, 1956.
Towne, Charles Wayland, and Edward Norris Wentworth. *Shepherd's Empire*. Norman: University of Oklahoma Press, 1945.

Tucker, David M. *Kitchen Gardening in America: A History*. Ames: Iowa State University Press, 1993.

Tull, Jethro. *The Horse Hoeing Husbandry*. Introduction by William Cobbett. 1731–41. Reprint, London: John M. Cobbett, 1822.

Turner, Arthur W., and Elmer J. Johnson. *Machines for the Farm, Ranch, and Plantation*. New York: McGraw-Hill, 1948.

Turner, Frederick Jackson. "The Significance of the Frontier in American History." In *The Turner Thesis Concerning the Role of the Frontier in American History*, rev. ed., edited by George Rogers Taylor, pp. 1–18. Boston: D. C. Heath, 1956.

U.S. Department of Agriculture. *[Annual] Report, 1875*. Washington, D.C.: Government Printing Office, 1876.

——. *[Annual] Report, 1888*. Washington, D.C.: Government Printing Office, 1889.

——. *Annual Reports of the Department of Agriculture, 1905*. Washington, D.C.: Government Printing Office, 1905.

——. *Annual Reports of the Department of Agriculture, 1910*. Washington, D.C.: Government Printing Office, 1911.

——. *Annual Reports of the Department of Agriculture, 1920*. Washington, D.C.: Government Printing Office, 1921.

——. *Annual Reports of the Department of Agriculture, 1921*. Washington, D.C.: Government Printing Office, 1922.

——. *Annual Report of the Secretary of Agriculture, 1925*. Washington, D.C.: Government Printing Office, 1925.

——. *Crops in Peace and War: Yearbook of Agriculture, 1950–1951*. Washington, D.C.: Government Printing Office, [1952].

——. *Grass: Yearbook of Agriculture, 1948*. Washington, D.C.: Government Printing Office, 1948.

——. *Science in Farming: Yearbook of Agriculture, 1943–1947*. Washington, D.C.: Government Printing Office, 1947.

——. *Trees: Yearbook of Agriculture, 1949*. Washington, D.C.: Government Printing Office, 1949.

——. *Yearbook of the Department of Agriculture, 1896*. Washington, D.C.: Government Printing Office, 1897.

——. Agricultural Research Service. *Common Weeds of the United States*. 1970. Reprint, New York: Dover, 1971.

——. Forest Service. *Range Plant Handbook*. 1937. Reprint, New York: Dover, 1988.

——. Forest Service. *The Western Range: Letter from the Secretary of Agriculture*. Edited by H. A. Wallace. Washington, D.C.: Government Printing Office, 1936.

——. Forest Service. Division of Range Research. "The History of Western Range Research." *Agricultural History* 18 (April 1944): 127–43.

Vaihinger, H. *The Philosophy of "As If."* London: Routledge & Kegan Paul, 1924.

Voigt, William, Jr. *Public Grazing Lands: Use and Misuse by Industry and Government*. New Brunswick, N.J.: Rutgers University Press, 1976.

Vrooman, Carl. "Grain Farming in the Corn Belt." U.S. Department of Agriculture Farmers' Bulletin #704, 1916.

Watson, A. K., and A. J. Renney. "The Biology of Canadian Weeds, [part] 6: *Centaurea diffusa* and *C. maculosa*." *Canadian Journal of Plant Science* 54 (October 1974): 687–701.

Watson, Alan, ed. *Leafy Spurge*. Weed Science Society of America Monograph Series #3. Champaign, Ill., 1985.

Watts, Lyle. "U.S. Forest Service." In *Fifty Years of Forestry in the U.S.A.*, edited by Robert K. Winters, pp. 165–91. Washington, D.C.: Society of American Foresters, 1950.

Way, Ruth, and Margaret Simmons. *A Geography of Spain and Portugal*. London: Methuen, 1962.

Webb, Walter Prescott. *The Great Plains*. New York: Grosset & Dunlap, 1933.

Wellman, Paul. *The Trampling Herd*. 1939. Reprint, Garden City, N.Y.: Doubleday, 1961.

Wentworth, Edward Norris. *America's Sheep Trails: History, Personalities*. Ames: Iowa State University Press, 1948.

West, Terry. "USDA Forest Service Management of the National Grasslands." *Agricultural History* 64 (Spring 1990): 86–98.

Westbrooks, Randy. "A Commentary on New Weeds in the United States." *Weed Technology* 5 (January–March 1991): 232–37.

Westover, H. L., and George A. Rogler. "Crested Wheatgrass." U.S. Department of Agriculture leaflet #104, 1934.

Wheeler, W. A. *Forage and Pasture Crops: A Handbook of Information about the Grasses and Legumes Grown for Forage in the United States*. New York: D. Van Nostrand, 1950.

Wheelhouse, Frances. *Digging Stick to Rotary Hoe: Men and Machines in Rural Australia*. Melbourne: Cassell Australia, 1966.

White, Hayden. *Metahistory: The Historical Imagination in Nineteenth-Century Europe*. Baltimore: Johns Hopkins University Press, 1973.

White, Lynn. *Medieval Technology and Social Change*. Oxford: Clarendon, 1962.

White, Richard. "Environmental History, Ecology, and Meaning." *Journal of American History* 76 (March 1990): 1111–16.

Whitson, Tom D., ed. *Weeds of the West*. Jackson, Wyo.: Western Society of Weed Science, 1991.

Wickson, E. J. "Irrigation in Field and Garden." U.S. Department of Agriculture Farmers' Bulletin #138, 1901.

Widmer, Jack. *Practical Animal Husbandry*. New York: Charles Scribner's Sons, 1949.

Widtsoe, John A. *Dry-Farming: A System of Agriculture for Countries under a Low Rainfall*. 1911. Reprint, New York: Macmillan, 1919.

Wilson, Gilbert. *Buffalo Bird Woman's Garden: Agriculture of the Hidatsa Indians*. St. Paul: Minnesota Historical Society Press, 1987.

Winters, Robert K. "The First Half Century." In *Fifty Years of Forestry in the U.S.A.*, edited by Robert K. Winters, pp. 1–29. Washington, D.C.: Society of American Foresters, 1950.

Wise, Gene. " 'Paradigm Dramas' in American Studies: A Cultural and Institutional History of the Movement." *American Quarterly* 31 (Fall 1979): 293–337.

Woods, John. "Enemy Fire!" *American Forests* 49 (April 1943): 232–58.

Wooten, H. H., and C. P. Barnes. "A Billion Acres of Grasslands." In U.S. Department of Agriculture, *Grass: Yearbook of Agriculture, 1948*, pp. 25–34. Washington, D.C.: Government Printing Office, 1948.

Worster, Donald. *Dust Bowl: The Southern Plains in the 1930s*. New York: Oxford University Press, 1979.

——. *Nature's Economy: A History of Ecological Ideas*. 1977. Reprint, Cambridge: Cambridge University Press, 1985.

——. *Rivers of Empire: Water, Aridity, and the Growth of the American West*. New York: Pantheon, 1985.
——. "Seeing Beyond Culture." *Journal of American History* 76 (March 1990): 1142–47.
——. "Transformations of the Earth: Toward an Agroecological Perspective in History." *Journal of American History* 76 (March 1990): 1087–1106.
——. *Under Western Skies: Nature and History in the American West*. New York: Oxford University Press, 1992.
——. *The Wealth of Nature: Environmental History and the Ecological Imagination*. New York: Oxford University Press, 1993.
Wright, Benjamin F., Jr. "Political Institutions and the Frontier." In *The Turner Thesis Concerning the Role of the Frontier in American History*, edited by George Rogers Taylor, pp. 34–42. Boston: D. C. Heath, 1956.
Young, James A. "Tumbleweed." *Scientific American* 264 (March 1991): 82–87.
Young, James A., R. A. Evans, R. E. Eckert, Jr., and B. L. Kiowa. "Cheatgrass." *Rangelands* 9 (1987): 266–70.
Zahn, Curtis. "The San Diego Fires . . . an Inquest." *American Forests* 50 (April 1944): 161–63.
Zinn, Howard. *A People's History of the United States*. New York: Harper Collins, 1980.

Index

Absarokas, xiv
Acomas, 59
Adirondacks, 25
Advertising Council, 42
Africa, 62
African Americans: facilitating fire fighting, 31; as settlers, 54; agricultural colleges for, 57; women, 71
Agricultural colleges, 57, 93, 151
Agricultural experiment stations, 57, 63, 93, 115, 131
Agricultural science, 2, 64, 70, 151–52, 153, 154
Agriculture: and colonization, x, 1, 57, 61, 66, 67, 78, 80, 91, 98, 144–45; productivity of, 2–3, 65, 74, 114, 140, 142, 151; and civilization, 3, 82, 93; etymology, 4; in Europe, 4–5, 50–54, 57, 60, 73; and improvement, 5, 16, 57–58, 61, 74–78, 79, 82, 83, 84, 118; as socially stratifying, 50–51, 73–74, 142, 145; and militarism, 52, 140; as commercial industry, 53, 55, 65, 69, 114, 117, 142; in Philippines, 57–58, 76; scientific, 63–64, 115; in South, 67; in East, 67, 76, 77; alternative, 70, 73, 153–54; scale of, 73, 74, 77, 114; "primitive," 75–76; sedentary, 79, 82
Agriculture, western: images of, 5, 7–9, 70–71; on prairie, 54, 77; scale of, 55–57, 59–60, 64, 65, 77, 82, 86, 136; on Plains, 56, 62–65, 67, 82, 97, 107–8; northwestern, 97
Agronomy, 63, 95
Agrostology, 80, 96–97, 152
Alaska, 37, 38, 39, 40, 42
Alderson, Nannie, x
Alfalfa, 62, 97–98, 114, 127, 142, 152
Allen, E. T., 30
Allen, Shirley, 20–21
Allotment, 58, 150
American Association for the Advancement of Science, 25
American Fat Stock Show, 86
American Forestry Congress, 43
American Hereford Cattle Breeders' Association, 86
American Service of Supply: Forestry Division of, 33
American Steel and Wire Company, 137
American studies, 5, 7, 8–9
Anglo-Americans, 84, 91, 110, 142, 143; as settlers, 59; as farm laborers, 69. *See also* European Americans
Anguses, 92
Animal husbandry, 80, 93–94, 95, 98, 152
Animal products, 94, 98, 102, 103, 104; beef, 82, 85, 86; hides, 82, 87; wool, 82, 87, 88, 90, 91, 92; mutton, 82, 88, 91, 92; dairy, 85, 86; as crops, 90, 106
Animals, 50, 51, 75, 111, 119, 126; traction, 66, 72, 73, 75, 135; wild, 81, 110; value of, 112. *See also* Livestock
Anishinabe, 122, 125, 126, 134
Anzaldúa, Gloria, x
Apaches, 59
Arapahoes, 59
Ard, 50, 60, 74. *See also* Plow
Ardrey, Robert, 76, 77
Argentina, 88, 89
Aridity, 59, 61–62, 64, 66, 67, 79, 107, 111, 112, 127, 128, 142, 143
Arizona, 68, 69, 88, 100, 131, 142, 143
Arizona Wool Growers Association, 100
Asia, 62, 66, 124, 127
Asian Americans: as farm workers, 69
Australia, 62, 76, 78, 88
Aztecs, 51

Bailey, Liberty Hyde, 116
Barbed wire, 61, 81, 83, 138
Beal, W. H., 152

Beard, Charles, 150
Beattie, W. R., 152
Benally, Capiton, 109–10
Bennett, Hugh, 110
Bergson, Henri, 10
Billington, Ray Allen, 82
Biodiversity, 4
Bison, 81, 82, 84, 87, 111
Bluegrass, 96, 97, 132
Bolley, H. L., 137
Bookchin, Murray, 13, 18
Botanists: as plant "explorers," 62, 97
Botany, 24, 45, 95, 96, 97, 101, 115, 120, 139
Bowman, J. H., 90–91
Bracero program, 69
Braudel, Fernand, 52
Brazil, 45, 88, 89
Britain, 18, 38, 83, 85, 90
British Columbia, 38
Brome, 62; downy (*Bromus tectorum*), 15, 99, 128, 129–31, 133, 134, 136; smooth, 97, 99
Buffalo. *See* Bison
Bureau of Animal Industry, 91, 111
Bureau of Indian Affairs, 57, 111
Bureau of Land Management, 105
Burgoyne, John, 125
Burlington Railroad, 63
Butler, Ovid, 40, 41

California, 41, 42, 44, 54, 62, 68, 69, 73, 81, 85, 88, 99, 113, 125, 131, 132, 142
Campbell, Hardy, 62
Canada, 81, 83, 88, 124, 125, 127
Canada thistle (*Cirsium arvense*), xiv, 124–26, 129, 133, 134–35, 139
Capitalism, 11, 111
Caribbean Sea, 78, 84
Carlisle Indian School, 60
Carson, Kit, 90
Cattle, 109, 120, 134, 136; Texan, 80, 81, 83–84; value of, 80, 84, 86, 95; markets for, 81, 83; Spanish, 81, 84–85; Eastern, 83, 84, 86; improvement of, 83, 84, 86–87, 89–90, 95, 107, 118; European, 85; bulls, 85, 86, 87, 89, 90, 92, 93; in Southwest, 86; Mexican, 88; compared to sheep, 88, 89; cows, 89, 90, 92, 93
Central America, 45, 62
Charlemagne, 51, 52
Cherokees, 54, 58, 122, 147
Cheyennes, 59
Chicago, 69, 83, 86, 137
Chickasaws, 54, 58
Chile, 88
China, 66, 67
Choctaws, 54, 58
Churros, 87–88, 90
Civilian Conservation Corps (CCC): and military mobilization, 35–36
Civil War, 54, 58, 61, 81, 84, 141
Clapp, Earle, 39
Clastres, Pierre, 74
Clay, Henry, 86
Clements, Frederic, 101–2, 140–41
Clover, 62, 120; alsike, 97, 98; red, 97, 98
Cochrane, Timothy, 31
Coleman, Norman, 95
Collier, John, 91, 109, 110
Colonization: etymology, 4, 5; in Europe, 52. *See also* Agriculture: and colonization
Colorado, 65, 68, 83, 85, 88, 89, 132
Comanches, 59
Congress of Industrial Organizations (CIC), 43
Connecticut, 134
Conquest, 2, 8, 18, 123, 152, 154
Conrad, Joseph, 150, 152
Cooley, W. E., 121
Copernicus, Nicholas, 14
Corriedales, 92
Cotswolds, 88, 90
Couchgrass (*Agropyron repens*), 129, 131, 132
Cowan, Charles, 31–32
Cowboys, 81, 95
Crabgrass (*Digitaria sanguinalis*), 129, 131, 132
Creeks, 54, 58
Crested wheatgrass (*Agropyron cristatum*), 129, 132–33

Cronon, William, 9
Crops: as food, 2; commercial production of, 3, 4, 64, 66, 81, 153; Spanish, 59; exploration for, 62–63, 124, 132. *See also* Food; Grain cultivation
Crosby, Alfred, 75, 113–14, 119, 121
Cuba, 45
Cultural studies, 3
Culture: etymology, 74–75. *See also* Nature; Society
Curti, Merle, 149

Dakotas (people), 141
Dakota Territory, 60, 62
Dalrymple, Oliver, 55
Darlington, William, 113–14, 115, 122, 124, 125, 140
Darwin, Charles, 10, 143
Dawes General Allotment Act, 58
Deere, John, 54
Deforestation, 17, 18–19
Deleuze, Gilles, 18, 23, 144, 147
Deloria, Vine, Jr., 17
Dendrology, 23, 24, 45, 80, 152. *See also* Forestry
Depression, 64–65, 105, 130
Derrida, Jacques, 5
Desert Land Act, 66
Deterritorialization, 18, 22, 23, 28, 40, 81, 82
Division of Agrostology, 96
Division of Botany, 62, 96
Division of Forestry, 25, 26, 100
Dobie, J. Frank, 107
Domestication, x, 3, 50, 73, 77–78, 96, 99, 109, 117, 118, 120, 132, 136, 145
Drinnon, Richard, 9, 10
Drought. *See* Aridity
Dry farming, 62–65, 78
Dust Bowl, 64–66, 97, 104, 105
Dutton, Walt, 104

Ecologists, 65
Ecology, 95, 101, 119
Egleston, Nathaniel, 25
Egypt, 66, 76
Emergency Appropriations Act, 104
Emerson, George, 25
England, 20, 88
Environment: adaptation to, 106–8, 111–12
Environmental justice, 154
Environmental regulation, 154
Epistemology: agricultural, 46, 154
Erosion, 128, 130, 133, 154
Europe, 3, 66, 87, 101, 120, 124, 127, 132, 133, 148, 150
European Americans, 3, 113; settlers, 4, 29, 120, 125, 147; farmers, 58, 60, 66, 82, 135; women, 70, 71; historians, 84, 123; invaders, 113, 140; explorers, 120–21, 122
Europeans, 3; medicinal tradition of, 123

Farm Credit Administration, 65
Faulkner, Edward, 75–76
Fechner, Robert, 36
Federal Bureau of Investigation (FBI), 42
Federal Reclamation Act, 66
Fernow, Bernhard, 24, 25, 26
Fiction: philosophy of, 10
Filaree (*Erodium cicutarium*), 124–43, 145
Fire, 27–30, 102; prevention and control of, 19, 30–33, 35, 40–41, 47; incendiary, 37, 40–42
Fire fighting: as military endeavor, 31–33, 43–44
Fires, historic: Hinckley (1894), 29; Peshtigo (1871), 29; Big Burn (1910), 30–32; Mann Gulch (1949), 31
Fire stories, 31
Fite, Gilbert, 82
Flagg, James Montgomery, 40–41, 42
Food: production, x, 50, 53, 65, 98, 114, 119, 153, 154; crops, 2, 53, 72; gathered wild, 51; commercial fruit and vegetables, 67–68, 70, 73; preservation of, 69; value of, 114
Food aid, 3, 66
Forage: improvement of, 94, 98, 99; productivity of, 97. *See also* Grass
Forage plants, 4, 97, 101, 111, 131, 132; drought-resistant, 62, 97; as crops, 81, 94, 98, 99, 100, 102, 104, 106, 107, 125; introduction of, 96, 97–98, 99, 129, 130,

131, 132; palatability of, 99, 101, 102, 130, 131, 132, 133, 134
Foresters, 3, 18, 22, 39, 47, 105
Forest Products Laboratory, 34, 38
Forestry, 23, 45, 80, 103; history of, 20, 21, 25–26, 45–46; in Europe, 20, 25; textbooks, 20–21, 24, 30; Cornell school of, 25; and the state, 25, 27–28; Yale school of, 26, 45; and private/public boundaries, 30–31, 47; and military mobilization, 36, 44. *See also* Foresters; Forest Service
Forests, 3, 50, 80, 103, 106; conservation of, 3, 20, 25, 27, 32, 35, 37–38, 42, 43, 46, 47, 111; military involvement in, 3, 32–34; as origin of civilization, 17–18, 78, 79–80; aboriginal use of, 18; as "virgin," 18, 19, 77; of Lake States, 18, 23, 29, 45; European, 18, 38, 40, 42; northwestern old-growth, 19, 23, 27, 30, 37–40, 43; policy for, 20, 28; management of, 20, 28, 32, 47; proliferation and coordination of control over, 20, 31, 45–46, 47, 80; and development of the state, 21; liquidation and exploitation of, 21–22, 23, 24, 25, 30, 39, 40, 45, 47; southern, 23, 37; reserves and National Forests, 23, 40, 47, 103–4, 105; management of versus management of trees, 23–24, 47; western, 24, 25, 33, 36–37; legislation for, 26–27, 32, 103; clear-cutting of, 27; eastern, 27, 29; reforestation of, 27, 35; as permanent, 29, 47; and military mobilization, 36–40, 47; Canadian, 40; as military targets, 40–41; of tropical Western Hemisphere, 44–45
Forest Service, 19, 20, 25, 26, 27, 32–34, 43, 100, 101, 103, 104, 105; relationship with logging companies, 22, 28, 30; Branch of Research, 34; information disseminated by, 34, 103; relationship with military, 35, 37, 39–40, 41, 42
Forney, D. R., 121
Forsling, Clarence, 105
Forster, E. M., 150
Foucault, Michel, 5, 13, 14
Foy, C. L., 121
France, 25, 38, 88
Frankfurt School, 6, 10
Franklin, Benjamin, 61
Frontier. *See* West
Frye, Northrop, 9

Gamboa, Erasmo, 69
Gardening: as women's work, 53, 70; relative to field agriculture, 53, 70, 72, 73–74. *See also* Horticulture; Women
Gardens, 64, 67, 70, 71, 72, 116, 120, 136, 152–53
Geertz, Clifford, 10
Gender, 3, 50, 92–93
General Land Office, 105
Georgia, Ada, 115, 116, 117, 131–32
Germans, 40, 140
Germany, 25, 29, 38, 132, 134, 148
Goats, 109, 111
Goodnight, Charles, 86
Government Printing Office, 152
Grain cultivation, 50–51, 52, 125, 135, 137, 144; axiomatic to agriculture, 50; associated with origin of civilization, 50–51; privileged over food production, 50–51, 54, 70; resulting in scarcity, 52–53
Granger, C. M., 40
Grass: as hay, 4, 94, 95–96, 98, 125, 131, 132, 133; tame, 4, 77, 94–95, 96, 99; wild, 77–78, 94, 96, 98, 99, 118; perennial, 80, 81, 102, 130, 131; native, 97, 98, 99, 106, 130; introduced, 97, 98–99, 106, 118; annual, 102; weedy, 129–33. *See also* Forage plants
Grasslands: and civilization, 79–80; agriculturalization of, 80; conservation of, 106. *See also* Range
Graves, Henry S., 26, 33
Gray, Asa, 25
Grazing, 4, 61, 94–95, 98, 101, 102, 127, 131, 133; on public lands, 80, 81–82, 100, 103, 105, 106; overgrazing, 80, 84, 99, 102, 110, 114, 128, 129, 130; on private lands, 80, 104, 106; regulation of, 100, 103, 104–6, 109–11

Grazing Service, 104–5
Greathouse, Charles, 152
Greece, 17–18, 19, 97
Greeley, William, 20, 23, 27, 28, 30, 31, 33, 35, 44
Green Revolution, 3, 78
Guattari, Félix, 18, 23, 144, 147

Hallyn, Fernand, 13–14, 15
Hargreaves, Mary, 62
Harrison, Robert Pogue, 17
Hatch Act, 57
Hays, Samuel, 46
Hegel, Wilhelm Friedrich, 10
Herbicides, xiv, 126, 127, 129, 137; 2,4-D, 115, 139–40, 141, 142; Sinox, 138–39; military uses of, 140
Herefords, 84, 85–86, 88, 89
Hess, Robert, 45
Hidatsas, 51, 59
Hilgard, Eugene, 25
Hirohito (emperor of Japan), 41
Historiography, x, 3, 5–6, 8, 9, 10, 12, 13, 14–16, 107
History: of American civilization, 2, 147–48; of West, 2, 147–50; intellectual, 3, 5; agricultural, 5; ecological, 5; environmental, 5, 6, 10, 12; natural, 6, 10, 107, 108; causality of, 6–8, 11; materialist versus idealist, 6–10; as progress, 15, 16, 108, 143, 153; of the forest, 19; colonial, 121; professional, 147; as science, 149, 150
Hitler, Adolf, 36, 116
Hohokam, 107
Holbrook, Stewart, 31, 39, 42
Holder, Preston, 59
Homestead Act, xiv, 7, 54–55, 57
Homesteads: size of, 54–55, 81; grazing, 61, 81; irrigated, 66, 81
Hopis, 59
Horses, 66, 107, 109, 111, 128, 135
Horticulture, 68, 70
Hough, Franklin B., 25
Hudson's Bay Company, 45
Humanities, 2, 12
Hundredth meridian, 7, 61
Hungary, 99
Hunting and gathering, 79
Hurt, R. Douglas, 65

Idaho, 31, 60, 62, 88, 125, 126, 132, 134
Immigrants, 73; as firefighters, 31; European, 54, 122, 127, 132, 134, 150; Mennonite, 62, 126; as gardeners, 73
Immigration: of people, 114; of plants, 114, 120
Imperial archive, 151–53
Incas, 51
Incendiary attacks: American, 29, 42; Japanese, 42
India, 67
Indian Reorganization Act, 110
Indian Service, 91, 109
Indian territory, 59
Industrial Workers of the World (IWW), 34–35
Insects, 119, 126, 141, 144
Iowa, 115, 127, 130, 137
Iron Age, 51
Iroquois, 147
Irrigation, 60, 62, 66–68, 73, 94, 95–96, 98, 106, 129, 152; and hydraulic society, 66–67, 107
Ise, John, 30
Italy, 18

Jameson, Fredric, 10
Japan, 29, 67, 141
Japanese, 40, 140
Jardine, James T., 100–101, 102
Jardine, William, 64
Jefferson, Thomas, 1
Johnson, Martin, 109
Johnsongrass (*Sorghum halapense*), 129, 131

Kansas, 54, 65, 83, 96, 126, 135
Kant, Immanuel, 10
Kepler, Johannes, 13
Kiowas, 59
Kipling, Rudyard, 16, 149
Knapweeds (*Centaurea* spp.), 127–29, 139
Knowledge, 15; production of, 2–3, 20, 103,

105, 143; gathering, 45; indigenous systems of, 121; limits to, 153; reclaiming, 153–54
Kochia (*Kochia scoparia*), 129, 133, 134
Kolodny, Annette, 5, 7, 8, 9, 10, 11, 70
Koyré, Alexander, 13
Kraus, E. J., 140
Kuklick, Bruce, 8, 9

Labor: organized, 34–35, 43, 46; corvée, 66, 67; contract, 68–69; migrant, 69; value of, 135
Lacan, Jacques, 14
Lagunas, 59
Land, 108, 154; use, 2, 107, 111; clearing, 3, 21, 22, 28, 50; as "virgin," 7, 8, 70–71, 74, 75; grants, 22; law, 22, 23, 111; value of, 56, 134, 136
Land-grant colleges, 7
Landlords, 51, 52
Leafy spurge (*Euphorbia esula*), 45, 124, 126–27, 129, 133
Lévy-Bruhl, Lucien, 10
Lewis, R. W. B., 9
Liebig, Justus, 137
Limerick, Patricia Nelson, 55
Lincoln, Abraham, 57
Linnaeus, Carolus (Carl von Linné), xiv, 120, 125
Little, Charles, 49
Livestock, 65, 80, 111, 115, 128, 130; wild or scrub, 4, 80, 84, 87, 89–90, 118; breeding, 4, 80, 86, 87, 89–93, 109; crops for, 52; industry, 62, 78, 91, 96, 98, 99, 106, 108, 125, 130, 136; quality of, 80, 87–88, 90, 92, 96, 99–100, 103; aspects under scientific control, 81, 89, 92–94, 112; bodies of, 83, 92–93, 94, 108–9; Spanish, compared to northern European, 84, 88; established in many colonies, 88–89; projection of social categories on, 92–93. *See also* Animals
Loggers, 34, 39, 42, 43, 46; "gyppos," 39
Logging, 3, 30; migratory, 18, 22–23, 28, 80, 82; technologies of, 22, 28, 37, 42, 43; regulation of, 28, 37–38, 39, 45; by military, 33; for war mobilization, 33, 37, 38–39, 42. *See also* Forests
Longhorns, 81, 84–87, 98, 118
Lorentz, Pare, 65
Loyal Legion of Loggers and Lumbermen (4L), 34–35
Lumber and Timber Producers' Defense Committee, 37
Lumbermen: and timber companies, 18, 19, 22, 24, 28, 29, 32, 34–35, 39, 46

McCarran, Patrick, 105, 111
McEntee, James, 36
Mack, Richard, 136
MacLean, Norman, 31
Maize, 51, 59, 132
Malin, James, 65
Mandans, 59
Manifest destiny, 2, 49, 142
Manitoba, 56, 126
Market gardens. *See* Truck farms
Marx, Leo, 9, 10
Maryland, 140
Massachusetts, 25, 126
Massumi, Brian, 147
Meadow fescue, 96
Mechanization, 53–54, 57, 61, 65, 69, 77, 135
Men, 51, 74; as firefighters, 32, 43; as farmers, 59, 70, 72, 74, 75; Native American, 60, 61; as farm laborers, 69
Merchant, Carolyn, 5, 10–13, 14
Merck, George W., 140
Merinos, 87, 88, 90, 91
Metallurgy, 51, 52
Mexican Americans/Mexicans: as farm laborers, 68–69
Mexico, 42, 44, 45, 67, 81, 85, 86, 88, 89, 102, 113, 140
Michigan, 29, 42
Mies, Maria, 50
Miller, Perry, 9
Miller, T. L., 86
Minnesota, 29, 55, 56, 60, 126, 127
Minuteman statue (Concord, Mass.), 49
Missionaries: Spanish, 58–59, 88; French, 124

Missouri, 128
Mitchell, J. W., 140
Mohegans, 122, 126
Monoculture, 70, 113, 134
Montagnais, 126
Montana, 31, 83, 88, 89, 95, 125, 126, 127, 132, 134
Morrill Land Grant Act, 57
Muenscher, Walter Conrad, 116, 128
Mumford, Lewis, 13

Nabhan, Gary, 153
National Industrial Act, 104
National parks, 40
Native Americans, 29, 66, 67, 113, 122, 123, 124, 140; as farmers, 4, 49, 54, 59, 120, 141, 153; migratory, 19, 23, 61; of Plains, 28, 59, 61, 81, 107, 108, 111; reservations; 35, 58, 60–61, 110, 112, 125, 144, 150; women, 49; agriculturalization of, 49, 57–61, 74, 78, 125; removal of, 54, 55, 141; wars against, 58, 149–50; self-determination of, 58, 150; as hunters and gatherers, 59, 76, 144; as laborers, 60; in boarding schools, 60–61; as stock raisers, 88, 90–92, 109–11, 112; knowledge systems of, 121; plant use of, 125; persistence of, 141–42; return of remains to, 153
Naturalization, 124; of plants, 113, 123; of people, 123
Nature, 2, 6, 82, 108, 153; intertwined with culture, x, 2, 4, 6, 10–12, 15, 16, 50, 74–75, 78, 123, 144, 152; scientific understandings of, 10–11; as machine, 11–12; as female, 15, 74, 116–17; etymology, 74–75; as farmer, 116–17; as separate from society, 121. *See also* Society; Women
Navajos (Diné), 59, 90–92, 109–11, 112
Nebraska, 86, 101, 126, 134–35
Nevada, 88, 105, 132
New Criticism, 8
New Deal, 35, 91, 104
New Mexico, 85, 88, 131, 132
Newton, Isaac, 57, 76
New York, 25
New Zealand, 88
Nietzsche, Friedrich, 1, 5, 14–15, 79
Noble, Marguerite, 142
Nomadism, 23, 61
North Africa: as source of Spanish agricultural resources, 78, 84–85, 87, 97, 142, 143
North Dakota, 55, 62, 96, 126, 127, 128, 132, 137
Northern Pacific Railroad, 55, 56
Northwest, 34, 127, 128, 130, 134
Norwood, Vera, 71
Nostalgia, 17, 74, 153; imperialist, 74–75, 78, 91
Nutrition, 94; of animals, 80, 81, 94; of plants, 137

Office of Indian Affairs, 35
Oklahoma, 54, 65, 86
Ontario, 125
Oregon, 37, 38, 39, 62, 68, 100, 101, 125, 131, 132, 134, 142
Osages, 59

Pacific Islands, 62
Pammel, L. H., 115
Paraguay, 44
Parkman, Francis, 149
Park Service, 40
Pastoralism, 61, 79, 80, 83, 107, 111. *See also* Stockraising
Patriarchy, 143, 154
Pennsylvania, 60, 133
Persia, 97, 98
Pesticides, 154; DDT, 141
Philippines, 67, 76
Pimas, 59–60, 77
Pinchot, Gifford, 24, 26, 27, 100
Pisani, Donald, 69
Plains, 124, 128, 133
Plantain, broadleaved (*Plantago major*), 121–24, 125
Plants: value of, 112, 118; native and introduced, 113, 118, 141; ornamental, 114, 120, 129, 133; culinary, 114, 122, 132; medicinal, 114, 122–23, 126, 133, 134; poisonous, 115, 128; multiple uses of, 118, 133, 134; as escapes, 120; classification of, 120–21;

ceremonial, 122–23; interaction with people, 153. *See also* Forage plants; Grass; Weeds
Plant succession, 80, 81, 101, 102, 107, 119, 140–41, 143, 145; compared to progress of civilization, 101–2, 140–41, 143, 145
Plow, 49, 51, 60, 63, 74, 75, 76, 77, 141, 144; and civilization, 3; and social stratification, 49–51, 52, 74, 75, 78; changes in, 50–52, 54; gang, 55, 56; breaking, 56, 77; axiomatic to agriculture, 60–61; in dry farming, 62–63
Poland, 40, 132
Portmanteau biota, 114, 119
Potter, Albert, 100, 103
Powell, John Wesley, 6–7, 66, 107, 111
Prairie, 130, 144; as "virgin," 3
Progress, 3, 21, 57–58, 74, 76, 82, 83, 101–2, 107–8, 111, 140–41, 143, 151, 153
Property, 19, 52, 75, 81, 118
Public domain, 19, 22, 23, 25, 82, 83
Pulaski, Edward, 31
Pyne, Stephen, 28, 32, 41, 102

Race, 93
Railroads, 7, 18, 22, 29, 63, 75, 120, 124
Rambouillets, 88, 91, 92
Range, 4, 61, 124, 125, 128; agriculturalization of, 4, 95, 107, 111, 136; productivity of, 4, 95–96, 98–100, 102, 136; fencing of, 61, 83; revegetation of, 62, 102, 110; quality of, 80, 98, 99, 100, 101, 103–4, 106, 109, 111; closing of, 83; improvement of, 96, 99; carrying capacity of, 96, 99, 100, 102–3, 109; aspects under scientific control, 111–12
Range management, 80–81, 94, 95, 99, 100–101, 104–6, 131
Range Plant Handbook, The, 125, 128, 129, 134
Range reconnaissance, 101, 102
Range science, 95, 99, 100, 102, 103, 104, 105–6
Reclamation, 66
Record, Samuel, 45
Redtop, 97, 98, 132
Regionalism, 7, 107–8, 111
Resettlement Administration, 65
Reterritorialization, 19, 52
Rhizome, 144–45
Richards, Thomas, 151, 152
Ridley, Henry, 120
Riordan, D. M., 90
Robbins, Wilfred, 138, 139
Roman Empire, 17–18, 19, 51, 87, 97, 122
Romneys, 91
Roessel, Ruth, 109
Roosevelt, Franklin Delano, 35
Rosaldo, Renato, 74
Rowley, William, 105
Russia, 62, 76, 89, 97, 126, 132
Russian thistle (*Salsola kali*), 128–29

Sahlins, Marshall, 49, 52–53
Sampson, Arthur, 94, 100–101, 102, 103
Sandoz, Mari, 84, 87
Santa Fe Railroad, 63
Sargent, C. S., 24
Schiller, Friedrich von, 10
Schopenhauer, Arthur, 10
Science, x; critique of, 2, 10–12, 150–51; and language, 13–15
Scribner, F. Lamson, 96
Sears, Paul, 65
Seminoles, 54, 58
Settlement, 19, 23, 26, 28–29, 57, 58, 79, 82, 136, 144, 149; Euro-American, 1, 22, 55, 58, 101, 108; Spanish, 87, 102, 108, 113
Sheep, 80, 81, 107, 109, 118, 120, 127, 136; improvement of, 84, 87–89, 90–92, 107; African, 87; Spanish, 87–88; northern European, 88; rams and ewes, 90, 93
Shepherds, 81
Shiva, Vandana, 113
Shorthorns, 84, 85, 86, 89
Shropshires, 93
Siberia, 132
Signal Corps: Spruce Production Division of, 33
Silviculture, 23, 24, 46. *See also* Forestry
Slotkin, Richard, 9, 10
Smith, Henry Nash, 5, 7, 8, 9, 10, 11, 61–62

Smokey Bear, 42
Social ecology, 13
Social justice, 154
Society: hierarchical structures of, 2–3, 6, 11, 13, 16, 32–33, 46, 50–54, 67, 74, 110–11, 112, 142, 143, 145; and nature, 6, 13, 108; adaptation to environment, 111–12; egalitarian, 153–54
Society of American Foresters, 26
Soil conservation, 65, 133
Soil Conservation Service, 65, 91, 104, 105, 110–11
Soil Erosion Service, 65, 110
Soil science, 95
Soil scientists, 63
South Africa, 88–89
South America, 45, 62
South Carolina, 131
South Dakota, 55, 96, 126, 127, 132, 135, 137
Southern Pacific Railroad, 63
Spain, 18, 84–85, 87, 88, 97
Spanish settlers, 142, 143
Spencer, Edwin, 132
State: knowledge gathering of, 3; powers of, 3, 19–20, 35, 67, 144; development of, 18, 29, 46–47, 74, 80, 143–44, 145
Stein, Sara, 118–20
Stockraising, 59, 61; open-range, 79, 80–84, 89, 108; scientific, 79, 81; agriculturalization of, 79, 82–83; as commercial, 79, 91, 108, 110–11; on northern Plains, 83; in Texas, 83–85
Strong, Josiah, 7, 150
Subjectivity, 13
Sudworth, George, 24

Taft, William Howard, 31
Taylor, Edward, 104
Taylor Grazing Act, 104
Technics, 13, 50
Technology, 49; transfer outside United States, 44, 57, 76; of warfare, 51, 52, 140
Territory, 2, 3, 18, 22, 45–46, 52, 79, 80–81, 82, 114, 143
Texas, 54, 65, 68, 69, 81, 83–85, 132
Third World, 78
Timothy, 96, 97, 98, 120, 132
Tools, 2, 153, 154; of men, 3, 50, 51; hand, 3, 74, 135; of women, 50, 71, 76; for field agriculture, 53, 78; Spanish, 59; "digging sticks" as, 60, 75, 76; for gardening, 73
Törbel, 111
Tractors, 66, 136
Trees, 80, 106; as a crop, 3, 24, 47, 107; harvest, 19, 24; replanting, 19, 24, 44; consumption of, 20, 33, 44, 80; sustained yield of, 28, 38; Douglas fir, 34, 38, 39; Sitka spruce, 38; Noble fir, 38; Asian teak, 39. *See also* Forests
Truck farms, 67, 116. *See also* Gardens
Tull, Jethro, 53–54, 63, 64, 72, 135
Turkey, 97, 127, 131
Turner, Frederick Jackson, 7, 9, 15, 16, 79, 82, 107, 147–51, 153

Union Pacific Railroad, 63
U.S. Army, 39, 140
U.S. Cavalry, xiv
U.S. Department of Agriculture (USDA), xiv, 7, 19, 25, 35, 44, 49, 62–63, 104, 122, 124, 128, 130; research and publications of, 57, 72–73, 80, 91–92, 96, 98, 139, 151–52
U.S. Department of the Interior, 19, 35, 104
U.S. Department of Labor, 35
U.S. Patent Office, 132
U.S. War Department, 33, 34, 35, 36, 38, 40, 140
Uruguay, 88
Usufruct, x, 73
Utah, 88, 142

Vaihinger, Hans, 10
Veterinary science, 95
Vico, Giambattista, 14
Viticulture, 53

War Production Board, 40
War Research Service, 140
Washington, 31, 37, 38, 39, 40, 62, 68, 125, 126, 127, 130, 132, 134
Webb, Walter Prescott, 6–7, 76, 77, 84, 107–8

Weed control, 4, 124, 131; mechanical, 114–15, 135, 136, 139, 142; as eradication, 115, 124, 131, 132, 141; by hand, 115, 135, 136; chemical, 115, 136, 137–40; by grazing, 127; cost of, 135, 136; as soil sterilization, 137–38. *See also* Herbicides
Weed migration, 120, 121, 123, 124–25, 126; by crop seed, 124, 126, 127, 128, 130, 134–45, 137; in hay, 125, 126, 130; by animals, 126, 130; by escape, 129, 130, 133, 134
Weed names, 122, 125, 128, 130, 132, 133, 142
Weeds, 80, 81; endangering agricultural productivity, 4, 78, 114, 115–16, 140, 141, 143; anthropomorphic qualities of, 4, 119, 126–27, 129, 132, 134, 141, 144, 145; as pioneer species, 102, 119, 120, 141, 145; embodying relationship between plants and people, 113–14, 119, 124, 140, 142–43, 145; legislation regarding, 114, 134–35, 142; history of, 115, 120, 121, 139; identification of, 115, 120–21, 139, 144; in competition with people, 116–17; definition of, 117–18, 119, 120, 140, 143; as ambiguous category, 118, 129, 144–45; physical properties of, 118–19; growth promoted by people, 119, 120, 121, 134; propensity to get away, 121, 143–45; people as, 141–42
Weed science, 137, 139, 152
Weed specialists, 118, 137
West: conquest of, 1; as frontier, 1, 2, 7, 36, 79, 82, 107–8, 147; as geographical place, 1, 2, 7, 107–8; as wilderness, 2; as origin of American civilization, 2, 147–48; images of, 5, 7, 8–9, 70–71; history of, 7
Western Forestry and Conservation Association (WFCA), 30, 31
West Indies, 45, 87
Weyerhaeuser Company, 27
Wheat, 52, 55, 59–60, 65, 76, 98, 114, 132, 152; bonanza farming, 55–56, 68; prices of, 56, 64, 65, 66; milling of, 56, 135; drought-resistant, 62
Wheatgrass, 62, 102; slender, 96, 97; western, 96, 97; crested, 97; intermediate, 97
White, Hayden, 14–15
White, Lynn, 76
Widtsoe, John, 63–64
Wilbur, Charles Dana, 61
Wilderness, 19, 46, 50, 78, 147, 148, 152, 154; textual, 149
William of Normandy, 52
Wisconsin, 29, 34, 137
Women, 8, 12, 69, 71, 74; and nature, 11; as firefighters, 43; as farmers, 49, 50, 53, 141; Native American, 49, 60, 73; status relative to men, 50, 51, 59, 68; as farm workers, 68; Mexican, 68; Mexican American, 69–70; as gardeners, 70–71, 73; "culture" of, 71, 78; as farmers' wives, 72, 73, 74, 78
Wood products, 23, 28, 37, 38, 44; research in United States, 33, 34, 38, 44; research in Germany, 38
Wood technology: German, 40
World War I, 33–34, 38, 39, 43, 64, 69, 103, 130
World War II, xiv, 3, 23, 29, 34–35, 36, 37–44, 45, 65–66, 103, 105, 141
Worster, Donald, 5, 6–7, 9–10, 11, 13, 65, 102, 107, 111
Wounded Knee, 141, 150
Wyoming, 83, 88, 89, 95, 96, 125, 127, 132, 134

Yellow toadflax (*Linaria vulgaris*), 129, 133–34

Zunis, 59